I0468682

CONTENTS

ACKNOWLEDGEMENTS

The authors express their appreciation to Captain Allender, Commander of the Naval Air Station, Adak, Alaska, and to the numerous Public Works officers who provided untiring assistance to the field crews and made field work on Adak bearable. Thanks also for the help (and humor) of many military and civilian personnel on the island.

Special thanks to Terry AtienzaMoore for keeping the '82 field crew organized (and entertained) and for all her help; and thanks to Jack Neffew, Jeff Roquemore, Dawn Janney, Ron Clodt, and Carl Halsey for help in the field and to Carla Gerrard for help in the office.

And last but not least, thanks to Steve Bjornstad and Ben Swann for their excellent drafting of figures and to Carole Anderson for her editing and comments.

INTRODUCTION

Adak Naval Station, Adak, Alaska is located approximately 1,200 miles southwest of Anchorage and 700 miles southeast of Siberia, near the center of the Aleutian island chain. The Station is strategically important for patrol and surveillance of Arctic waters. On Adak, all electric power generation and space heating is done with JP-5 jet fuel, amounting to nearly 9 million gallons per year. This dependence on oil, coupled with the remoteness of the island, makes any form of indigenous power source attractive if it is exploitable commercially. The purpose of this report is to present the data, results, and interpretations of numerous geoscientific investigations concerning indigenous geothermal potential at the Naval Station (NAVSTA), Adak.

GEOGRAPHY

Adak Island is located near the middle of the Aleutian chain in the Andreanof Islands (Figure 1). The island is roughly 30 miles east to west, 20 miles north to south (289 square miles), has 212 miles of shoreline, and is the largest island in the central Aleutian chain.

Most of the island is mountainous and rugged, and it has an irregular shoreline marked by numerous bays and indentations (Figure 2). Southern Adak, while rugged, probably contains glacially dissected marine platforms and is pocked by numerous small stream-fed lakes (Reference 1). Yakak Peninsula, an area of extensive lowlands, probably resulted from preglacial wave planation modified by stream and glacial erosion. The eastern coast is characterized by steep bluffs and rocky cliffs. Mount Moffett rises nearly 3,900 ft on the northwest end of the island, while Mount Adagdak rises to over 2,000 ft on the northeast side. In between the two mountains lie the remnants of the Andrew Bay Volcano. Also prominent on the northern end of the island are the large freshwater lake, Andrew Lake; the saltwater embayment, Clam Lagoon; and the open-water indentations of Kuluk Bay and Sweeper Cove. Finger Bay, structurally controlled by the Finger Bay Fault, lies south of Sweeper Cove.

Adak has the maritime climate common to all the Aleutian Islands of persistently foggy, wet, and stormy weather. The cloud ceiling on the island can range from 1,000 to 3,000 ft, and precipitation occurs almost every day, normally in great quantity. Located within the North Pacific storm tract, the island is frequently buffeted by intense storms, which produce winds in excess of 50 knots (although this is just a little brisker than the normal, day-to-day wind, which ranges from 10 to 33 knots). However, because of the moderating nearby warm Japanese current, the average annual temperature of the island is cool but not severe. The average monthly temperatures of Adak range from 0°C in February to 11°C in August.

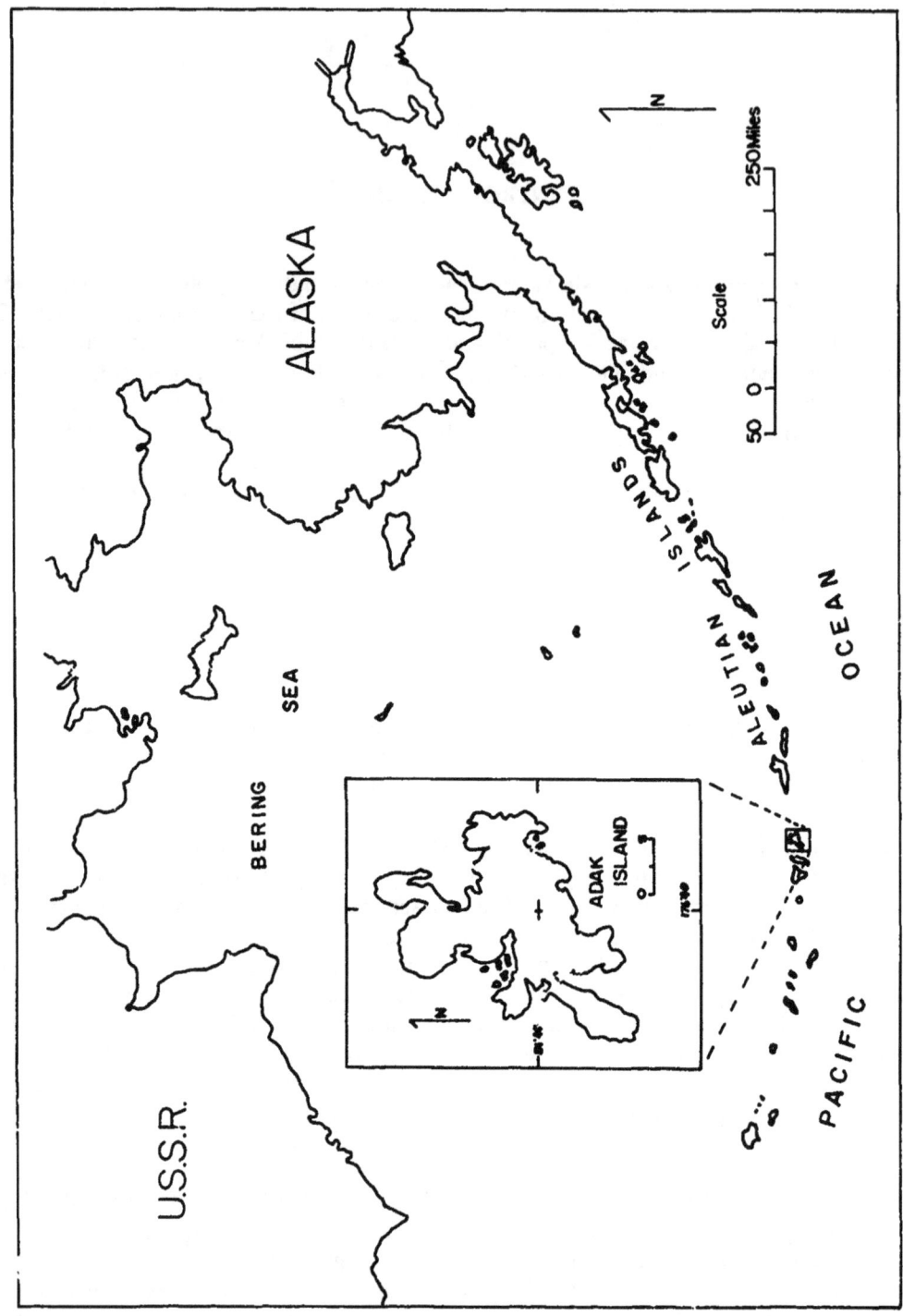

FIGURE 1. Location Map of Adak Island, Alaska.

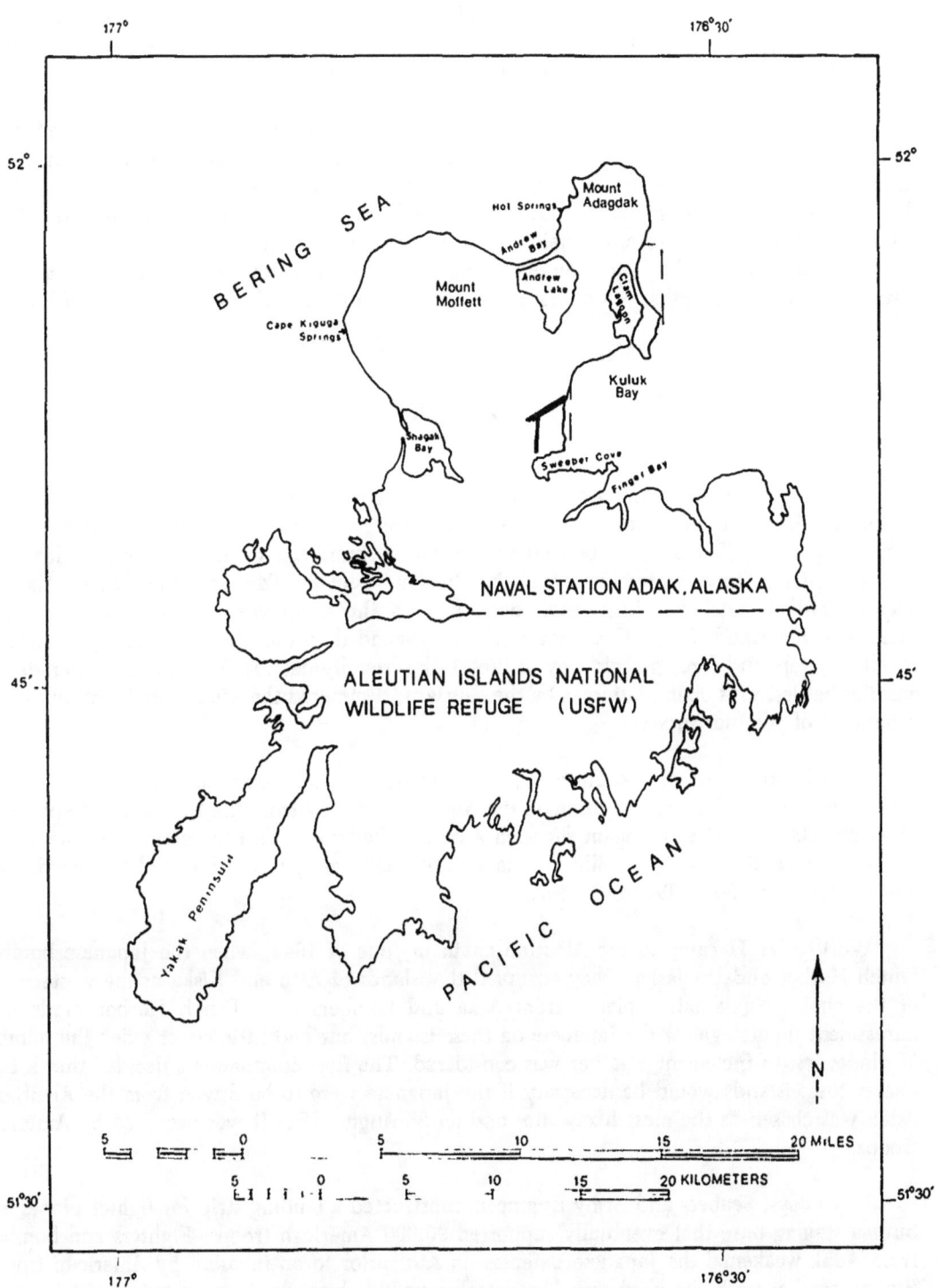

FIGURE 2. Geographic Map of Adak Island.

The volcanic nature of the Adak soil, coupled with the short growing season and low average summer temperature, retards tree and vegetable growth. On the other hand, tundra, a foot-high grass with a spongy base, grows well. During the summer the entire island is covered with a green carpet of tundra spotted with wildflowers. During the rest of the year, the tundra dies back to an all-encompassing and, some consider, a depressing brown.

Adak is the headquarters for the Aleutian Islands National Wildlife Refuge. The refuge of 2,720,235 acres comprises nearly 70 named islands of the Aleutian chain. The southern half of the island belongs to the refuge (see Figure 2). Birds are the most obvious feature of the refuge; the bald eagle is resident in substantial numbers. A large herd of caribou on Adak is the direct result of transplantation from the mainland and careful management. Also, the streams issuing from the island are used by large numbers of spawning salmon, to the point of overspawning, during the summer months. Sea otters are abundant in the coastal coves and in Clam Lagoon.

HISTORY

Adak is believed to have been first settled by the Aleuts, who supposedly crossed the prehistoric land bridge from Asia. The Aleuts were an easygoing, good-natured people; numbering from 20,000 to 25,000, they located in hundreds of island villages along the Aleutian chain. On 9 September 1741, the Russian vessel *St. Paul*, commanded by Captain Alexei Chirikof, anchored in a small open bay on the south coast of Adak and officially discovered the island. Later, Cossack fur traders descended on the Aleutian Islands, demanding tributes of furs and forcing their way of life on the friendly natives. The Aleuts scattered; and ensuing battles, plus open massacres by the Russians, decimated the Aleut population until only a handful of colonies survived.

Adak became part of the United States in 1867, when Secretary of State William Henry Seward persuaded Congress to purchase the Alaskan Territory from the Russians. The purchase at $7,200,000 seemed extravagant for such a barren wasteland, and the action became known as "Seward's Folly." Now, 7 million dollars would not even pay the fuel bill to provide heat and electricity to NAVSTA for 1 year.

World War II came to the Aleutian chain in June of 1942, when the Japanese bombed Dutch Harbor and Unalaska. They occupied the islands of Attu and Kiska at the western end of the chain. Navy patrol planes from Atka and bombers from Dutch Harbor made long harassment flights against the Japanese on these islands, but had little effect when the number of planes lost to inclement weather was considered. The fleet commanders decided that a base nearer these islands would be necessary if the Japanese were to be driven from the Aleutians. Adak was chosen as the most likely site, and on 28 August 1943 it was occupied by American troops.

In 10 days, Seabees and Army Engineers constructed a landing strip for fighter planes and built a staging base that eventually supported 90,000 American troops. Fighters and bombers from Adak weakened the Japanese defenses on Attu prior to an invasion by American troops. Bloody combat with the hard-core Japanese troops lasted nearly 2 weeks before the Japanese surrendered.

A joint Army-Navy team surveyed Adak Island in 1946 to establish a permanent Navy base. After the war, until 1948, Adak was operated by the Air Force but controlled by the Army. In July 1950 NAVSTA, Adak was commissioned. In 1951, the U.S. Naval Communications Station (COMSTA) was established; subsequently, in 1977, COMSTA was renamed the Naval Security Group Activity (NAVGRUSECACT). In 1971, the Fleet Air Alaska Command was transferred from Kodiak Island to Adak. The base now includes a naval facility, Marine battalion, Patrol Wings Pacific detachment, weather detachment, explosive ordnance detachment, and patrol squadron detachment. A seismology laboratory, maintained by the University of Colorado, and the wildlife refuge are tenant activities.

The mission of NAVSTA, Adak is to provide support for the antisubmarine warfare program and to conduct search and rescue operations in the North Pacific. Also NAVGRUSECACT serves as a key link in the Navy's communications network. The population of Adak, including NAVSTA, NAVGRUSECACT, the Naval Facility, the Alaska State School System, the scientific community, the U.S. Fish and Wildlife Service, and contractors, is between 4,500 and 5,000 military personnel, dependents, and civilians.

GEOSCIENTIFIC INVESTIGATIONS

GEOLOGY

Bradley provided the first significant geologic study of Adak Island when he published his "notes on geology" that he made while stationed on the island in the mid-1940s (Reference 2). While his study was not intensive, he did note the northward shift in volcanic activity, cited three periods of volcanism on the island, and commented on the glaciation and structure of the island. He did not, however, provide a geologic map.

Coats spent 25 days in the field in 1946 to prepare the geologic map most commonly accepted for northern Adak Island (Reference 3). This map is reproduced as Figure 3 in this report. Coats recognized two physiographic and geologic divisions on the mapped part of the island: (1) a southern area comprising folded, faulted, and intensely altered volcanic rocks of Paleozoic age that were deeply glaciated and then intruded by rocks of younger gabbroic and intermediate composition; and (2) a northern area dominated by three distinct volcanoes (Andrew Bay Volcano, Mount Moffett, and Mount Adagdak) of Tertiary or Quaternary age. Near the central part of the map are five volcanic domes of light-colored andesite porphyry that Coats feels are separate from the three basaltic volcanoes and probably are of early Tertiary age. The volcanoes have been trimmed by marine erosion and dissected by subaerial and ancient glacial erosion and have minor amounts of sedimentary rocks associated with them. The lowland areas of the island are covered by a blanket of ash from volcanoes on other, nearby islands.

An excellent summary of geology relating to a possible geothermal resource is an unpublished report by Miller and Smith (1977). Miller and Smith's work is sufficiently important that, rather than paraphrasing their findings in small sections, this report includes the entire summary as an appendix. Of particular interest are the sections on regional geologic setting, local geology, geochronology, and petrology that can be found in Appendix A.

7

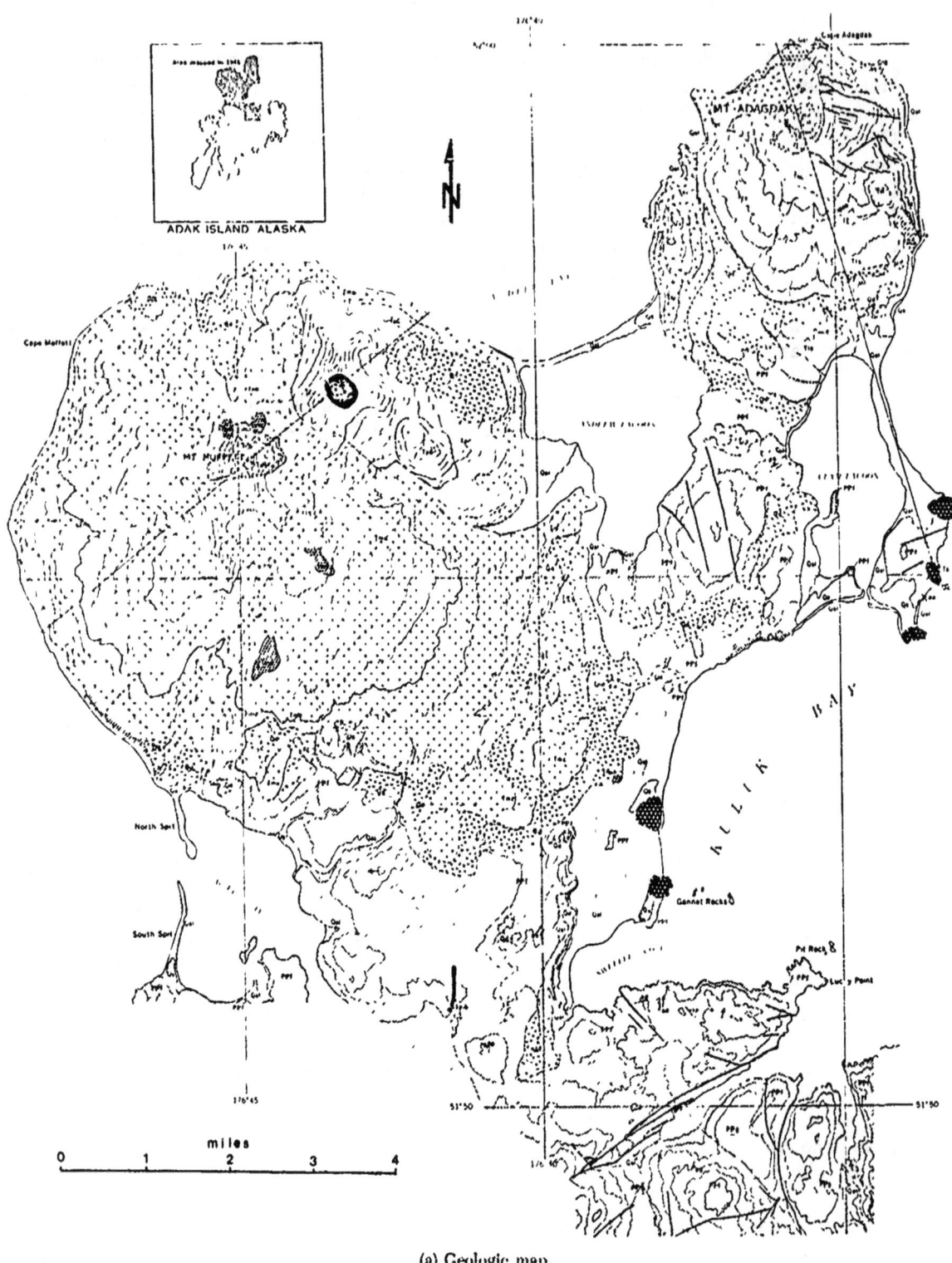

(a) Geologic map.

FIGURE 3. Geologic Map of Northern Adak Island (from Coats, 1956).

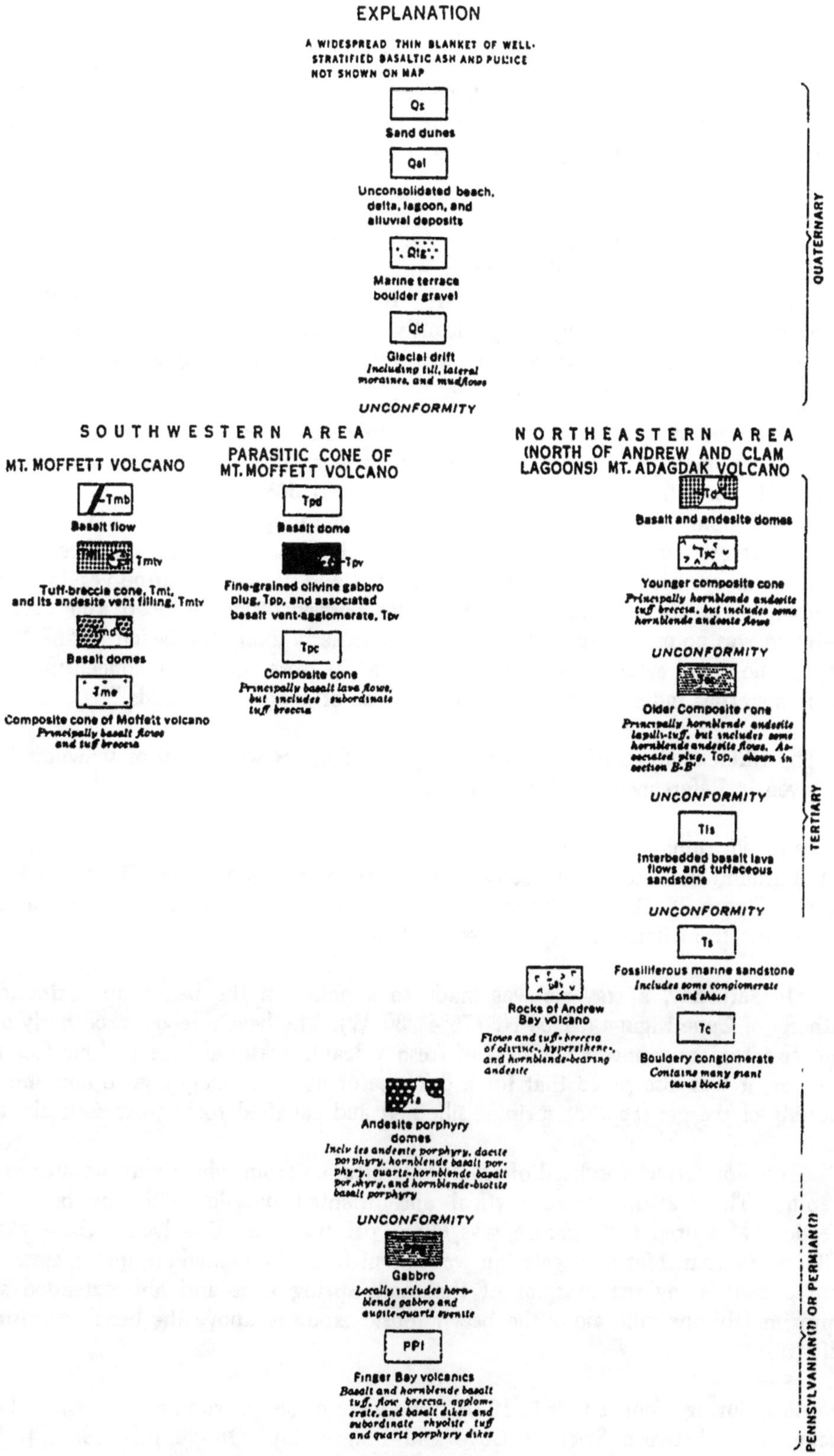

EXPLANATION

A WIDESPREAD THIN BLANKET OF WELL-
STRATIFIED BASALTIC ASH AND PUMICE
NOT SHOWN ON MAP

QUATERNARY

| Qs |
Sand dunes

| Qal |
Unconsolidated beach,
delta, lagoon, and
alluvial deposits

| Qtg |
Marine terrace
boulder gravel

| Qd |
Glacial drift
*Including till, lateral
moraines, and mudflows*

UNCONFORMITY

SOUTHWESTERN AREA

MT. MOFFETT VOLCANO

| Tmb |
Basalt flow

| Tmtv |
Tuff-breccia cone, Tmt,
and its andesite vent filling, Tmtv

| Tmd |
Basalt domes

| Tmc |
Composite cone of Moffett volcano
*Principally basalt flows
and tuff breccia*

**PARASITIC CONE OF
MT. MOFFETT VOLCANO**

| Tpd |
Basalt dome

| Tpv |
Fine-grained olivine gabbro
plug, Tpp, and associated
basalt vent-agglomerate, Tpv

| Tpc |
Composite cone
*Principally basalt lava flows,
but includes subordinate
tuff breccia*

NORTHEASTERN AREA
(NORTH OF ANDREW AND CLAM
LAGOONS) MT. ADAGDAK VOLCANO

TERTIARY

Basalt and andesite domes

Younger composite cone
*Principally hornblende andesite
tuff breccia, but includes some
hornblende andesite flows*

UNCONFORMITY

Older Composite cone
*Principally hornblende andesite
lapilli-tuff, but includes some
hornblende andesite flows. Associated plug, Top, shown in
section B-B'*

UNCONFORMITY

| Tls |
Interbedded basalt lava
flows and tuffaceous
sandstone

UNCONFORMITY

| Ts |
Fossiliferous marine sandstone
*Includes some conglomerate
and shale*

| Tc |
Bouldery conglomerate
*Contains many giant
talus blocks*

Rocks of Andrew
Bay volcano
*Flows and tuff-breccia
of olivine-, hypersthene-,
and hornblende-bearing
andesite*

Andesite porphyry
domes
*Includes andesite porphyry, dacite
porphyry, hornblende basalt porphyry, quartz-hornblende basalt
porphyry, and hornblende-biotite
basalt porphyry*

UNCONFORMITY

Gabbro
*Locally includes horn-
blende gabbro and
augite-quartz syenite*

| PPf |
Finger Bay volcanics
*Basalt and hornblende basalt
tuff, flow breccia, agglomerate, and basalt dikes and
subordinate rhyolite tuff
and quartz porphyry dikes*

PENNSYLVANIAN(?) OR PERMIAN(?)

(b) Legend.

FIGURE 3. (Contd.)

GEOCHEMISTRY OF HOT AND WARM SPRINGS

Miller and Smith (Appendix A) give a good description of the hot and warm springs on Adak Island. Another good description of the springs is given by James L. Moore and W. R. Tipton of the California Energy Company, Inc. (CECI). Moore and Tipton visited Adak in the summer of 1981 to evaluate the geothermal potential of the island. Moore's description of the Andrew Bay Hot Springs (see Figure 2) is quoted from Reference 4 with the author's permission:

"On Friday, July 31, I visited the hot spring located on the northwest coast of Mt. Adagdak, northerly off the NE corner of Andrew Lake (51°58′N, 176°38′W). This particular spring is already noted in the literature. However, it should be noted that the surface expression of the active spring is significantly different from the description in the literature. These surface changes can probably be attributed to an extremely severe storm that occurred during December of 1980.

"The large hot pool noted by others is gone, and the only geothermal evidence remaining is a series of small gas vents and hot water springs. The measured spring temperature was 155°F, pH was 5.5, field titration for CO_2 yielded 668 ppm and field titration for HCO_3 yielded 512 ppm. The springs are issuing from a group of vertical NW trending fractures extending across approximately 150 feet of basic igneous rock. One significant feature of the area near the springs was the abundance of dead and decaying marine vegetation; apparently the result of geothermal emissions. A similar concentration of dead and decaying marine vegetation was noted approximately one mile northerly along the beach (51°59′N, 176°37′W). Although no direct evidence of geothermal emissions were noted, the concentration of the dead marine materials indicated the probable presence of geothermal fluids."

The CECI analyses of the Andrew Bay Hot Springs water and of seawater from Andrew Bay given in Reference 4 are listed in Table 1.

Persistent rumors of hot springs on Mount Moffett led Moore in 1981 to explore the surface on and around that mountain. He discovered warm springs at Cape Kiguga on Mount Moffett (refer to Figure 2). His description of these springs follows. An analysis of water from these springs, also from Reference 4, is given in Table 1.

"On Saturday, a traverse was made to a point on the beach approximately one mile southerly of Cape Kiguga (51°54′N, 176°47′30″W). The beach deposits southerly of the subject point consisted of rounded boulders of fresh volcanic materials one to four feet in diameter. However, it was also noted that for a distance of approximately one to one and a half miles southerly of the gas seep/hot springs, silicified and calcified rocks were seen along the beach.

"The 'hot spring' consisted of fractured volcanics from which warm water and CO_2 were escaping. The fractures were vertical and oriented parallel with the beach in a S45°E direction. Measured temperature was 60°F, pH was 4.5. Dissolved CO_2 = 870 ppm with HCO_3 at 36 ppm. Marine vegetation was absent from the immediate spring areas. Iron springs were present along the margins of the main spring area and also extended southerly for approximately one mile along the beach. Basic exposure above the beach consisted of glacial drift (till)."

Also, during Moore's visit in 1981, there were persistent rumors of warm or hot springs on Lucky Point, between Sweeper Cove and Finger Bay. On 29 July 1981, J. Whelan and

TABLE 1. Chemical Analysis, Andrew Bay Hot Springs, Kiguga Warm Springs, and Andrew Bay Seawater.

Value measured	Andrew Bay Hot Springs			Kiguga Springs		Seawater CECI 3
	USGS 76AMm220[a]	USGS 76AMm221[a]	CECI 1	CECI 2	NAVY 1	
Dilution	?	?	1:7	1:6	...	...
pH	7.4	7.5	...	..	7.0	...
Specific conductance, µmho/cm	...	...	>10,000	3,900	...	>10,000
Temp. (meas.),°C	63	71	...	...	...	...
Constituent, ppm[b]						
Calcium	1500.	1300.	1300.	21.	13.	290.
Magnesium	70.	110	460.	38.	7.	340.
Sodium	6800.	6100.	5800.	500.	72.	6800.
Potassium	460.	380.	460.	38.	5.1	340.
Carbonate					0	
Bicarbonate	420.	430.			85.8	
Chloride			14000.	1100.	93.5	15000.
Sulfate	120.	330.	350.	200.	21.	2300.
Nitrate					4.0	
Fluoride	0.49	0.55	0.6	0.26	0.08	0.83
Iron			3.8	0.87	< 0.05	0.20
Manganese					< 0.01	
Arsenic					< 0.01	
Copper					< 0.01	
Zinc					0.01	
Mercury					< 0.0002	
Silica			192.	15.0	36	...
Aluminum					< 0.2	
Boron	87.	70.	89.0	0.39	0.06	3.5
Phosphate					0.2	
Bromine					< 0.2	
Lithium					< 0.01	
Nitrite					< 0.002	
Ammonia N			1.5	3.1		> 0.22
Ammonium					...[c]	
Ammonium (by calculation)[d]			1.9	4.0		> 0.26
Hydroxide					none	

[a] Refer to Appendix A.
[b] ppm = parts per million.
[c] Insufficient sample.
[d] Calculation of amount of ammonium expected at the time sample was taken in the field. Derived from the value of ammonium detected in the laboratory.

R. Clodt, of the Naval Weapons Center (NWC), China Lake, Calif., joined Moore in exploring the Lucky Point area. Moore's description is given below:

"On Wednesday, July 29, a traverse was made to Lucky Point (51°51'N, 176°35'W), a peninsular parcel of land between Sweeper Cove and Finger Bay. It was reported that hot springs were located somewhere along the north coastal portion of the point. The entire point area was examined for approximately the easterly-most mile of the peninsula and no evidence of active hydrothermal activity was defined. A small amount of surface discoloration and extremely minor $CaCO_3$ precipitation was noted on the rocks immediately above sea water level in a little cove about one mile west of the point (51°51'N, 176°36'30"W). This site is not believed to represent any 'real' surface manifestation of geothermal activity. No direct evidence of active geothermal activity was found anywhere on the peninsula and the reported 'hot spring' was in all likelihood one of the numerous very shallow tundra lakes that dot the peninsula's surface that had been warmed by the sun to a temperature that was pleasant to the 'skinny dippers'."

Reports of surfur-bearing fumaroles on the upper slopes of both Mount Moffett and Mount Adagdak were not evaluated in the field.

CHEMICAL GEOTHERMOMETRY

Hot and warm springwater geochemistry is valuable in evaluating geothermal resources. Calculations were made at the Andrew Bay Hot Springs and Kiguga Warm Springs to determine temperatures of various geothermometers. Results are given in Table 2.

TABLE 2. Chemical Geothermometers, Andrew Bay Hot Springs, Kiguga Warm Springs, and Andrew Bay Seawater.

Geother-mometer[a]	Andrew Bay Hot Springs			Kiguga Springs		Seawater CECI 3
	USGS 76AMm220[b]	USGS 76AMm221[b]	CECI 1	CECI 2	NAVY 1	
	Values in °C					
a	174	173	167	62	91	. . .
b	186	185	177	54	87	. . .
c	165	164	155	20	54	. . .
d	149	142	164	160	153	123
e	188	182	194	185	82	185
f	. . .	. . .	. . .	162	. . .	169
g	187	181	. . .	. . .	. . .	. . .

[a] Geothermometers:
  a—Quartz, steam flashing (Fournier, 1981, Reference 5).
  b—Quartz, conductive cooling (Fournier, 1981, Reference 5).
  c—Chalcedony, conductive cooling (Fournier, 1981, Reference 5).
  d—Sodium-potassium (Truesdell, 1976, Reference 6).
  e—Sodium-potassium-calcium (Fournier and Truesdell, 1973, Reference 7).
  f—Sodium-potassium-calcium-magnesium (Fournier and Potter, 1979, Reference 8).
  g—Sodium-potassium-calcium-carbonate (Paces, 1975, Reference 9).
[b] Refer to Appendix A.

Of interest is the general agreement between the silica, quartz conductive cooling temperature and the sodium-potassium-calcium temperatures for the Andrew Bay Hot Springs. Because of possible seawater encroachment, the writers believe the silica geothermometers to be most appropriate, with either the quartz conductive cooling or the chalcedony conductive cooling applying. Therefore, our best estimate of reservoir temperature for the Andrew Bay Hot Springs is 155 to 186°C. For the Kiguga Warm Springs the temperatures would range from 54 to 91°C.

The calculated temperatures stated above compare well with those published by Motyka (Reference 10). He calculated temperatures of 160 to 188°C for the deep reservoir temperature. Motyka also states (in a personal communication) that there is a magmatic influence on the hydrothermal system. He determined this influence by use of helium (^{3}He/^{4}He) and analyses of ^{13}C in fumarolic CO_2.

Because of high precipitation on Adak, it was decided to try "mixing models." The silica mixing model of Truesdell and Fournier is shown as Figure 4 (Reference 11). The Kiguga springs were used as groundwater; the Andrew Bay springs as the warm spring. Since the line through the Kiguga point and the Andrew Bay points does not intersect the quartz solubility line, the "no steam loss" model does not apply. Assuming that enthalpy in international table calories per gram has nearly a one-to-one correlation to temperature in this range of °C,[*] and

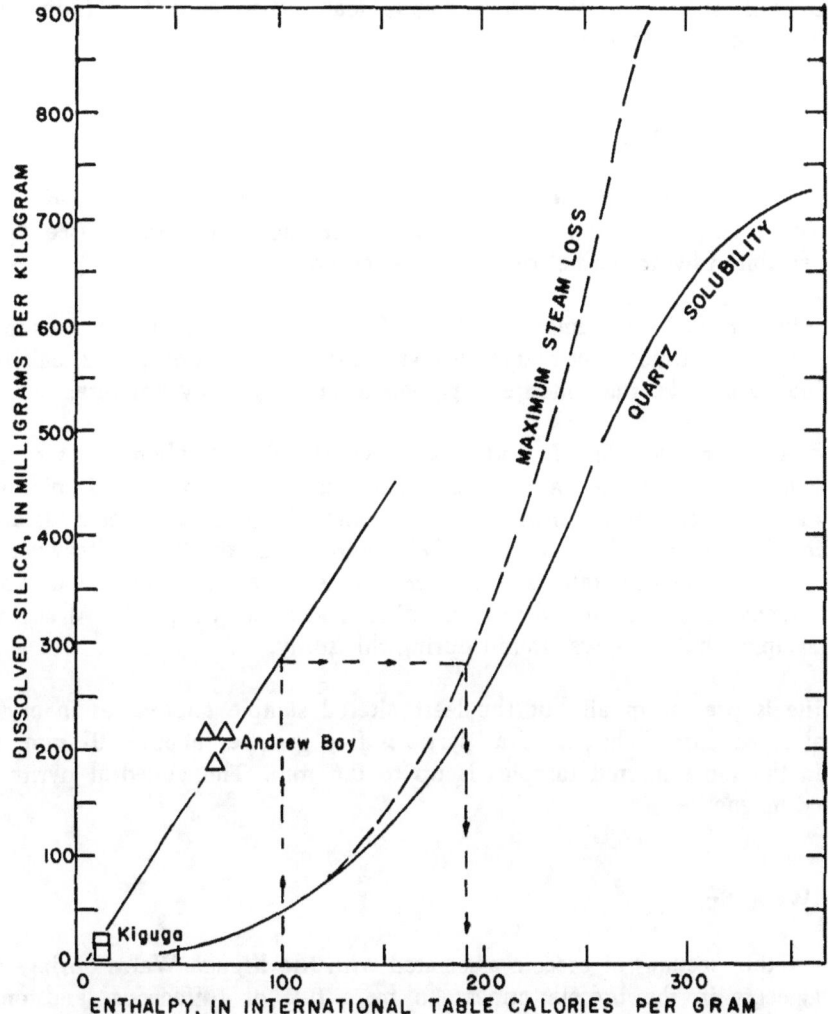

FIGURE 4. Silica Mixing Model of Truesdell and Fournier (1979) Applied to Geothermal Waters, Northern Adak Island.

[*] Refer to steam tables in a Handbook of Chemistry and Physics, for example.

assuming a steam loss at 100°C and one atmosphere pressure, the reservoir temperature would be 190°C. Based on equations from Fournier (Reference 12), the silica content of the reservoir would be about 220 ppm and there would have been a 21% steam loss. There are not enough data to apply the chloride mixing model.

HYDROTHERMAL ALTERATION

Geothermal solutions interact chemically with the rocks they traverse, depositing new minerals in fractures and replacing rock minerals. Opal, chalcedony, quartz, calcite, aragonite, pyrite, marcasite, sulfur, realgar, orpiment, and cinnabar are frequently deposited. Earlier rock-forming minerals may be changed to chlorite, epidote, serpentine, alunite, montmorillonite, kaolin family minerals, secondary biotite, K-feldspar, etc. Both the Andrew Bay Hot Springs and the Kiguga Warm Springs areas exhibit hydrothermal alteration. Samples collected by Brophy were studied petrographically (Reference 13). Individual thin-section descriptions are given in Appendix B.

Andrew Bay Hot Springs

Four thin sections from the Andrew Bay Hot Springs area were studied. Three of the thin sections were andesites and the fourth an andesite porphyry breccia (see Appendix A). All sections exhibited hydrothermal or deuteric alteration.

The fine-grained groundmass (0.03 to 0.5 mm) seen in the thin sections make up 60% of the rock and consist mostly of plagioclase with minor amounts of quartz, calcite, and epidote. In the breccia sample, the andesite fragments are cemented by limonite.

Phenocrysts are hornblende and plagioclase. Hornblende phenocrysts range in size from 0.6 to 3 mm; all have been at least partially altered to chlorite and epidote. In the more severely altered rocks the transformation is complete. Plagioclase phenocrysts are euhedral and range from 0.5 to 4 mm. They are severely fractured and exhibit oscillatory zoning. Often they have several optical orientations and overgrowths. They are calcic and when altered are partially replaced by calcite and sericite. Miller and Smith (Appendix A) also noted augite in similar samples, but none was found during this study.

Pyrite is present in all but the least altered sample and varies in crystal form from anhedral to euhedral. The anhedral pyrite is fine grained, about 0.01 mm, while euhedral pyrite in the more altered samples is up to 0.6 mm. The anhedral pyrite may represent sulfidized magnetite.

Kiguga Warm Springs

Three thin sections of breccia associated with the Kiguga Warm Springs were prepared. The fragments in the breccia appear to be a tuff or tuffaceous sandstone. Microscopic examination indicated a groundmass or matrix consisting of glass (0.01 mm including shards), quartz (0.05 to 0.15 mm), and K-feldspar. Microphenocrysts consist of euhedral K-feldspar ranging from 0.25 to 1 mm and making up about 25% of the rock; anhedral quartz averaging 0.25 mm, making up 30% of the rock; and a few percents of euhedral-zoned plagioclase ranging in size from 0.2 to 0.6 mm; thus, the fragments are rhyolite from an unknown source. The cement of the brecia is chalcedony.

The rock weathers tan, with areas containing pyrite going to brown because of limonitization. Hydrothermal alteration has sericitized the feldspars and probably some of the glass. Sericitization ranges from moderate to severe. Both euhedral and anhedral pyrite, in crystals and patches up to 0.2 mm, have been introduced. Quartz veinlets cut through the rock, and barite has possibly been introduced.

Of the two springs, the hydrothermal alteration of the Andrew Bay Hot Springs is both less intense and less apparent visually than that at Kiguga Warm Springs because the soft hydrothermal alteration minerals are easily eroded from the wave-cut terrace where they occur. The alteration of the Andrew Bay Hot Springs is essentially propylitic, with the introduction of pyrite and the conversion of magnetite to pyrite. This pyritic conversion probably accounts for a magnetic low seen by an aeromagnetic survey described in a later section. The alteration at Kiguga Warm Springs is more severe phyllitic (sericitic) alteration, with the introduction of quartz, chalcedony, and pyrite into the rock.

PREVIOUS GEOPHYSICAL STUDIES

Aeromagnetics

An aeromagnetic survey was flown over northern Adak Island in 1948, and the results were reported by Keller, Meuschke, and Alldredge (Reference 14). The survey was flown at an elevation of 5,000 ft (1.5 km) and a spacing of 1 mile (1.6 km). Observed data are shown on Figure 5, and a second derivative is shown on Figure 6.

On Figure 5, a large anomaly of approximately 700 gammas is seen south of Finger Bay and is ascribed by Keller to the large mass of gabbro in that area. Anomalous highs, associated with the volcanic domes on Mount Adagdak and the cone of olivine-basalt on Mount Moffett, are seen on both Figure 5 and Figure 6, as is the magnetic low over Kuluk Bay (although only slightly on the second derivative map). Keller points out that the low over Kuluk Bay may not represent a great thickness of sediments. The low may represent a function, characteristic of magnetics in the higher magnetic latitudes, where a well-defined low forms directly north of a magnetic high; e.g., the Finger Bay high. Also seen on the second derivative map is an unclosed low forming directly west of Mount Adagdak in the general vicinity of the Andrew Bay Hot Springs.

Microearthquakes I

From 22 October through 1 November 1974, more than 1000 km^2 were surveyed near Adak Island for microearthquakes. The objective was to detect and locate microearthquakes and thereby map tectonically active structures. Butler and Keller (1975, Reference 15) reported on the results that would apply to the evaluation of geothermal potential at Adak.

The Aleutian Arc is an active seismic province, with a large number of earthquakes occurring on or near the subduction or Benioff Zone of the islands' arc-trench system (see

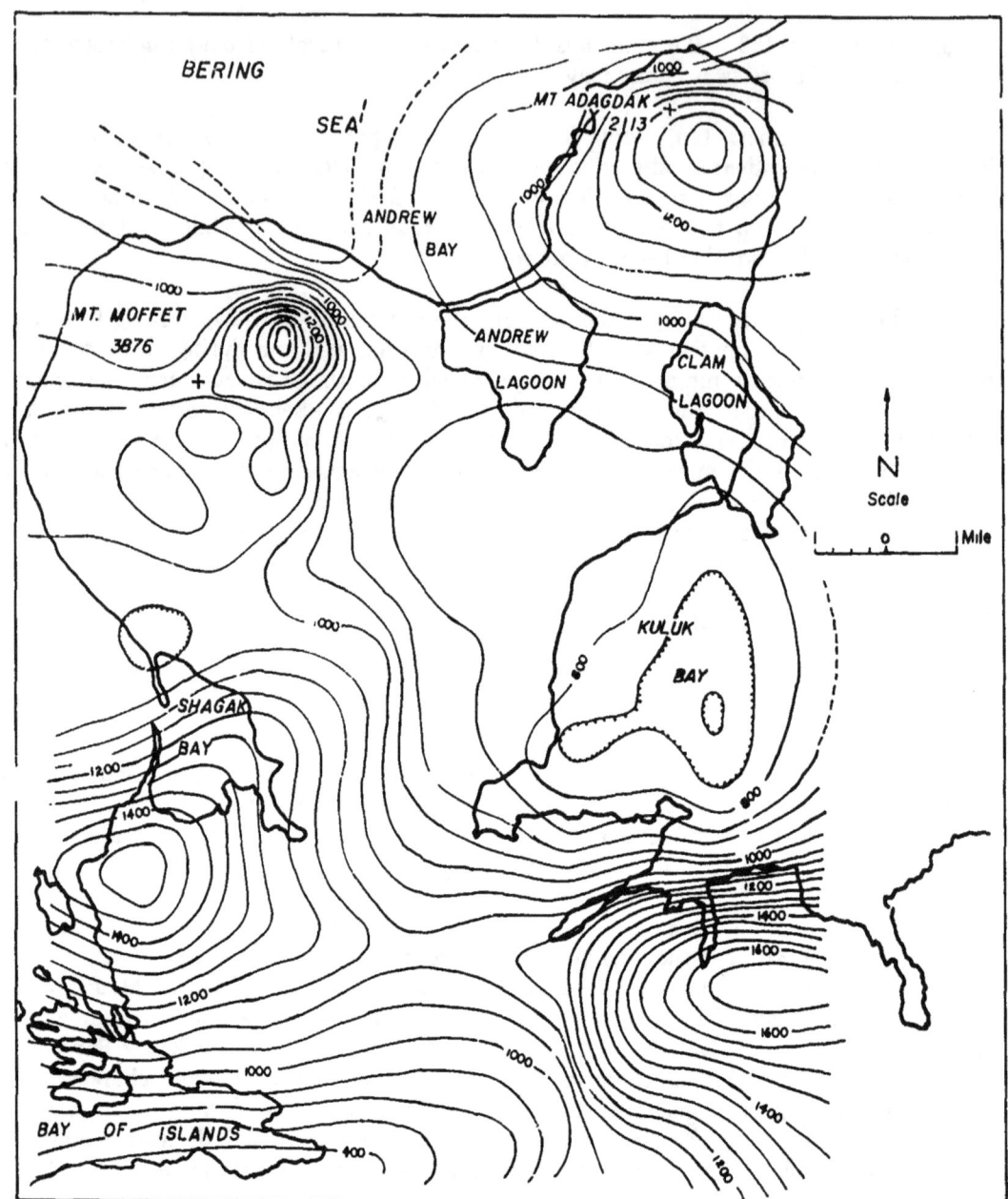

FIGURE 5. Observed Aeromagnetic Map of Northern Adak Island. Contour interval 50 gammas (from Keller et al, 1954).

Engdahl, 1975; Reference 16). In the Aleutian Islands, the North Pacific Plate is postulated as being moved in a northwesterly direction as it is subducted under the American Plate. This subduction zone strikes east-west south of Adak and dips northerly to a depth of 100 to 150 km beneath the island. Although the seismic activity associated with the seismic zone is too deep to be of interest in a microseismic survey, it is possible that the tectonic activity at depth is evidence of regional stress and that the stress pattern may influence surface faulting, with resultant seismicity.

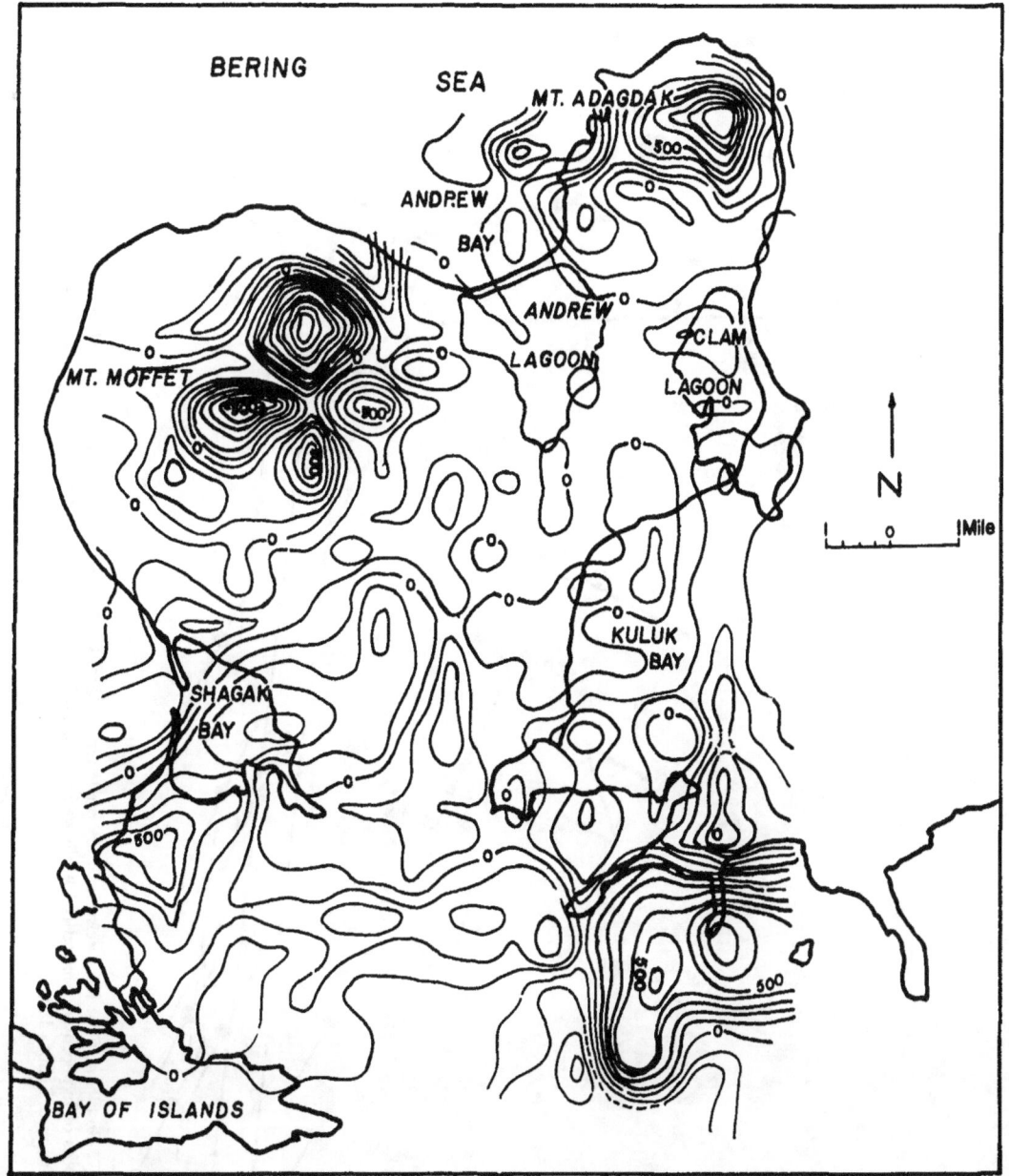

FIGURE 6. Second Vertical Derivative Map of Northern Adak Island. Contour interval 100 gammas/(0.5 mile) (from Keller et al. 1954).

Results from the survey indicated a seismically active zone near Andrew Bay striking through Mount Adagdak. Butler and Keller interpreted it to be a right-lateral, strike-slip fault with a small component of thrust (Reference 15). Allowing for uncertainties in the depth parameter from the velocity model they were using, Butler and Keller fitted a fault plane striking N60°E and dipping 70°NW (Figure 7). They also hypothesized the existence of a fractured or dilated region caused by this fault zone, located southwest of Mount Adagdak (centered along the spit north of Andrew Lagoon) and northeast of Mount Adagdak.

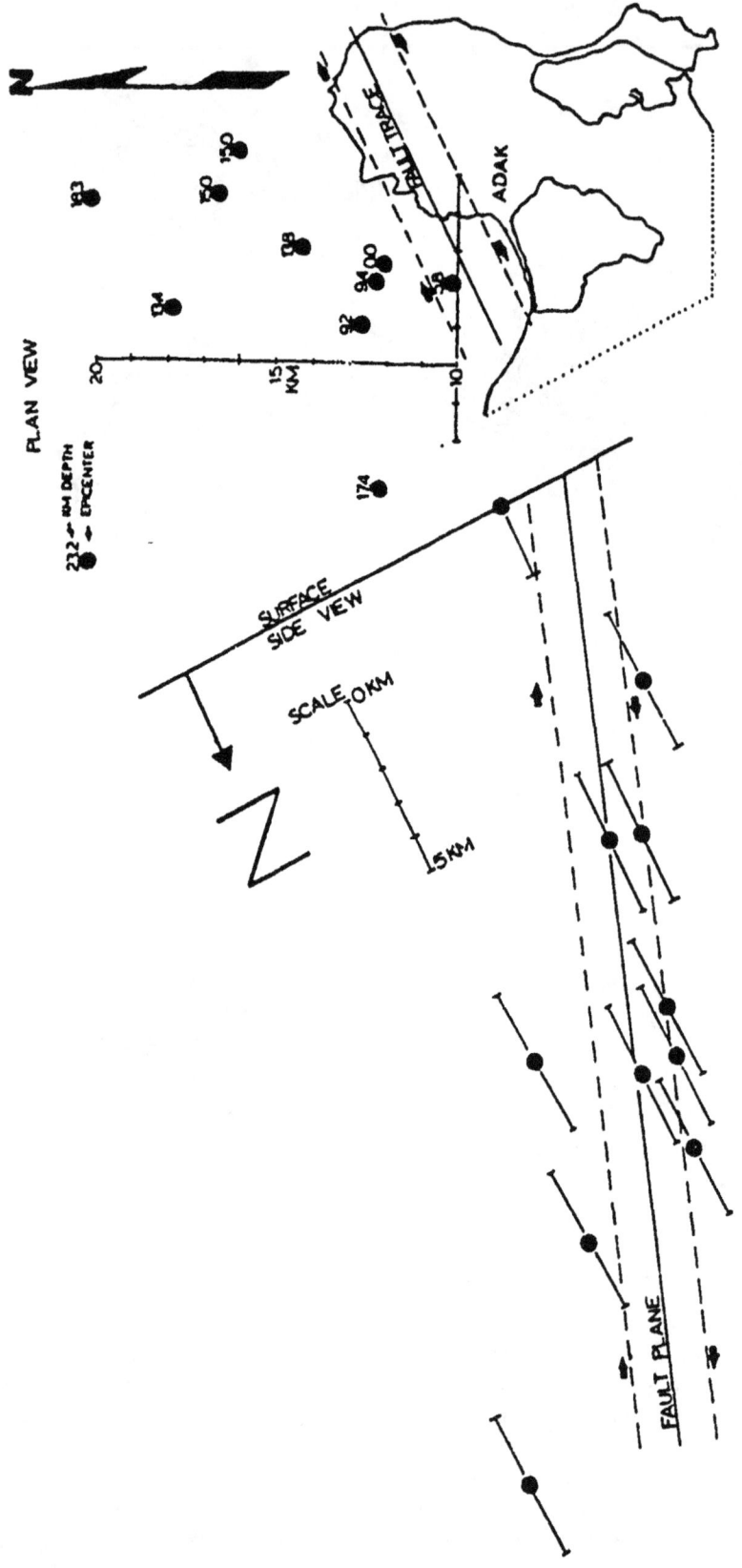

FIGURE 7. Results of First Microseismic Study on Mount Adagdak. Cross section constructed normal to strike of fault trace, depths of hypocenters plotted (with allowed error), and least-square-fit solution to fault plane drawn (modified from Butler and Keller, 1975).

18

Electrical Surveys I

Butler and Keller also reported on a resistivity survey carried out on Adak by the Colorado School of Mines in the fall of 1974 (Reference 15). The survey used the rotating dipole technique along the southwest flanks of Mount Adagdak. Because of weather conditions, relatively few measurements were made before the survey was terminated. Results show a high resistive layer close to the source and then decreasing with depth (Figure 8). At distances greater than 1 km the apparent resistivity values decrease to less than 20 ohm-meters. These low values are compatible with geothermal fluids at depth.

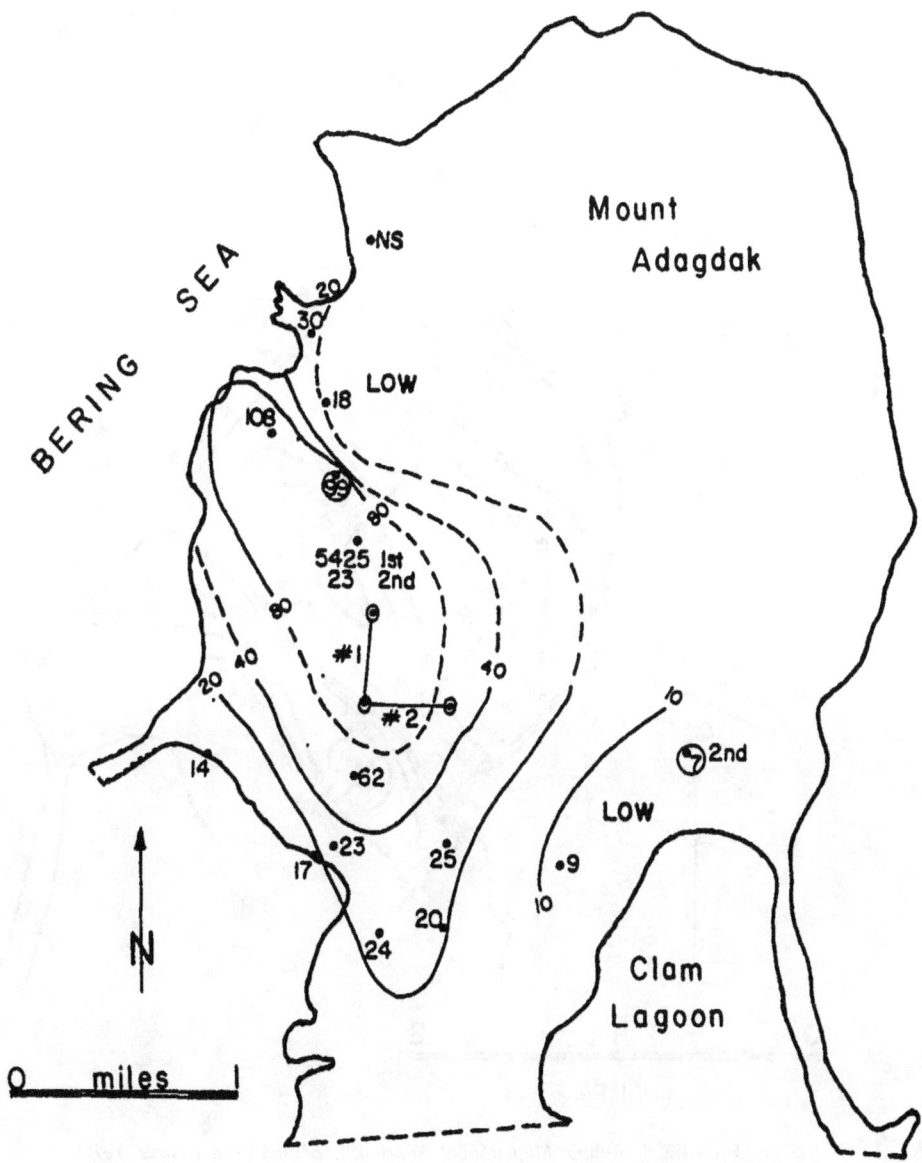

FIGURE 8. Apparent Resistivities Measured on Mount Adagdak in 1974 (modified from Butler and Keller, 1975).

Electrical Surveys II

With the above information in hand, NWC tasked the United States Geological Survey (USGS) to explore Adak Island for a geothermal resource using various geophysical techniques. The studies were carried out in July and August of 1976 and included, among other things, the use of audio-magnetotelluric, telluric, Geonics EM-16R, and self-potential electrical surveys. Unlike the survey performed by the Colorado School of Mines, these electrical studies used natural instead of induced signals. Hoover (Reference 17), in a preliminary report referred to as the Hoover report, described the results paraphrased as follows:

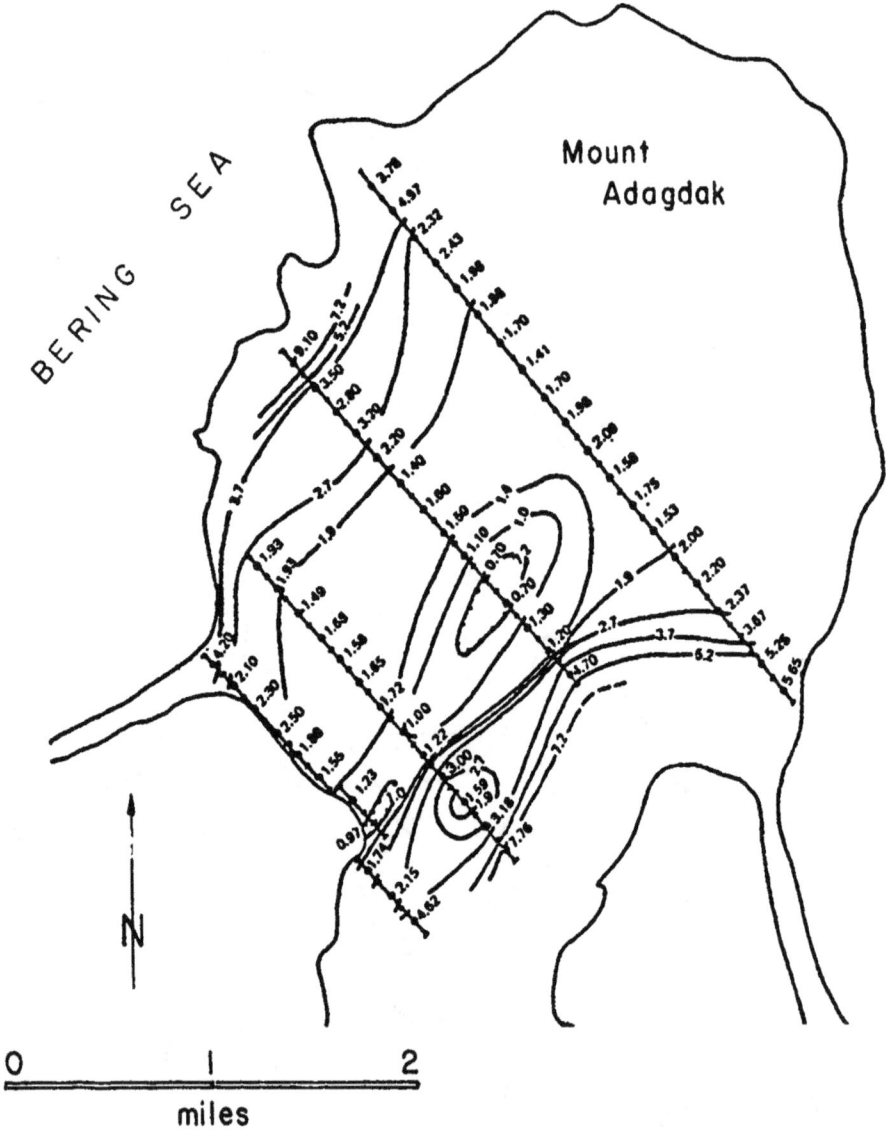

FIGURE 9. Telluric Map, Mount Adagdak (modified from Hoover, 1976).

Magnetotellurics. Preliminary results from this survey suggest that the northern portion of Adak Island has considerably more electrical conducting properties than the southern part. The change in intensity of the electric field variations, from Teardrop Lake (approximately 10 km or 6.2 miles south of Finger Bay) to the southwest of Mount Adagdak, is approximately 25:1. This implies a 625:1 change in resistivity. The polarization of these currents on the northern portion of the island is indicative of the presence of a northeasterly trending conductor in this region (especially near Mount Adagdak). The persistence of the polarization of currents in a northeasterly direction, using periods of up to 1,000 seconds, suggested to the USGS that the conductor is sizable.

Tellurics. Seven telluric traverses were attempted on the southwest flanks of Mount Adagdak; six of the traverses were parallel to each other in a N42°W direction; the seventh was orthogonal to the others.

The telluric voltage map is shown by Figure 9, using the six parallel lines for information. The small closed low on the southeast end of traverse 1 is in an area with a number of buildings and may represent buried conductors. However, a northeast trend is apparent in the data. The USGS feels this reflects the principal structural trends on the island. It suspects that a broad conductor exists beneath that part of Adagdak surveyed and that resistivity contrasts of about 100 to 1 are present between the edges of the volcano and the most conductive region in the center. Since the low resistivity anomaly has a short wavelength, it is probably shallow. The USGS concludes that this could be a fault zone that has channeled thermal solutions nearer to the surface.

Geonics EM-16R. This technique uses very low frequency radio transmitting stations for an energy source and directly measures ground resistivity in the frequency range of 15 to 25 Khz. The survey was taken in conjunction with the telluric traverses and was useful in helping to determine the upper value of resistivity on the audio-magnetotelluric soundings. Plotted values for this survey are shown on Figure 10.

Audio-magnetotelluric. This technique uses the orthogonal horizontal electric and magnetic fields caused by worldwide lightning storms. Measurements are made from a range of frequencies from 7.5 to 18,000 Hz, with the lower frequencies giving greater depth penetration.

Using the data gathered in the audio-magnetotelluric survey, one-dimensional models were computed. Although these models are considered only approximate, a very good conductor on the order of 1 ohm-meter was identified at a depth as shallow as 602 m.

Self-potential. Figure 11 shows the location and information gathered (in millivolts (mv)) from the self-potential (SP) survey done by the USGS on Mount Adagdak. Extremely large SP voltages are noted on the north part of the island along with large voltage gradients. However, these voltages are thought to be due to the topography effect characteristic of an SP survey performed in mountainous areas (Mount Adagdak in this case) and not to any geothermal system.

The USGS did locate what appeared to be a long wavelength anomaly in the area of Mount Adagdak. It discovered this possible anomaly by running a line northwest along the spit

21

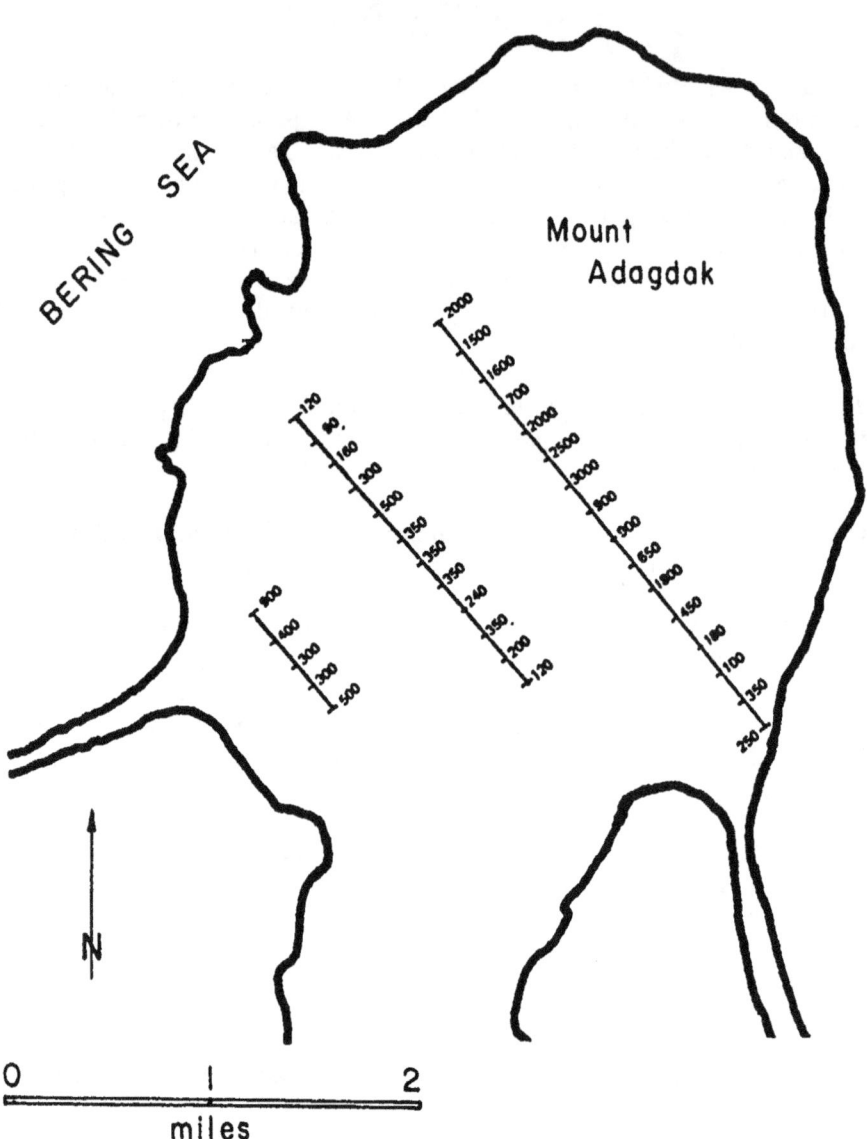

FIGURE 10. Apparent Resistivities Measured at 24 kHz (modified from Hoover, 1976).

between Andrew Lake and the Bering Sea. This traverse is nearly radial to Mount Adagdak, with less than 5 m variation along the line. The original intent of the line was to see if any expression of the fault identified by Butler and Keller (Reference 15) could be detected by using the SP method. The fault was not detected, either because the amplitude of the feature was too small or because the structure was farther west than the line extended. However, when the line was extended to the northeast, up the slope of Adagdak, four other data points were obtained where the line crossed previous traverses. There, the USGS detected a 200-mv anomaly of unsymmetrical shape. The USGS is quick to point out that these four points are at a higher elevation than the 'spit' line and that some topographic correction should be applied. However, the correction would only increase the amplitude of the long wavelength to perhaps 400 mv and reduce the asymmetry. The USGS concludes that while the SP data are very speculative, they do suggest a large, deep source below the center of Mount Adagdak with the correct signature for a geothermal system.

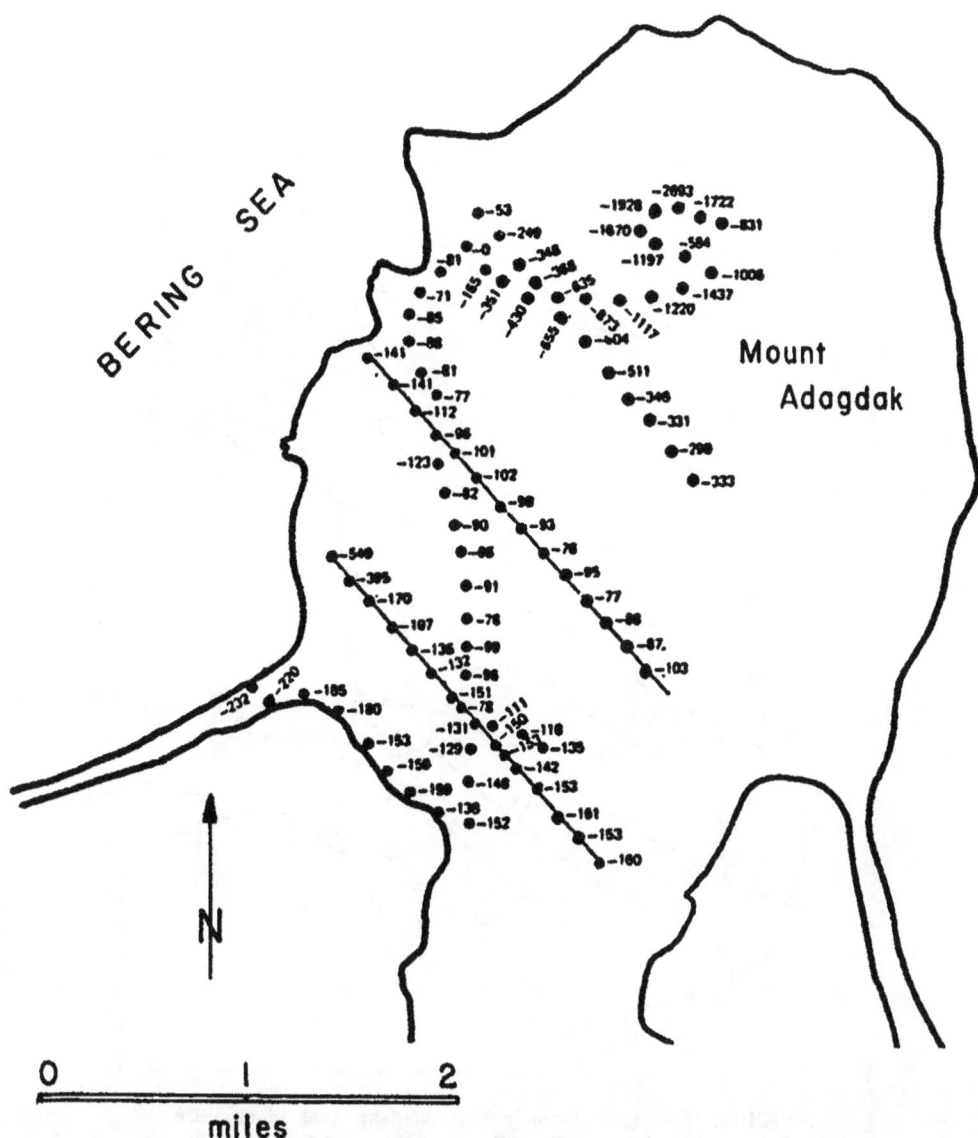

FIGURE 11. Self-Potential Values, Mount Adagdak (modified from Hoover, 1976). Values in millivolts.

Gravity I

The data obtained from 30 gravity stations established on the northern part of Adak were reported in the Hoover report. Elevation accuracies of the stations were considered to be within 1.5 m with the exception of one station. The survey used the Adak Base Station at the Naval Air Station as a reference for elevation control. Terrain corrections were made to a distance of 2.6 km. The Bouguer anomaly values are posted and contoured on Figure 12.

Figure 12 shows that the gravity decreases to the north at about 2 milligals* (mgals) per km. Near Mount Adagdak and to the south a local gravity low and perhaps a smaller gravity

* A unit of acceleration used with gravity measurements; 10^{-3} gal = 10^{-5} m/s^2.

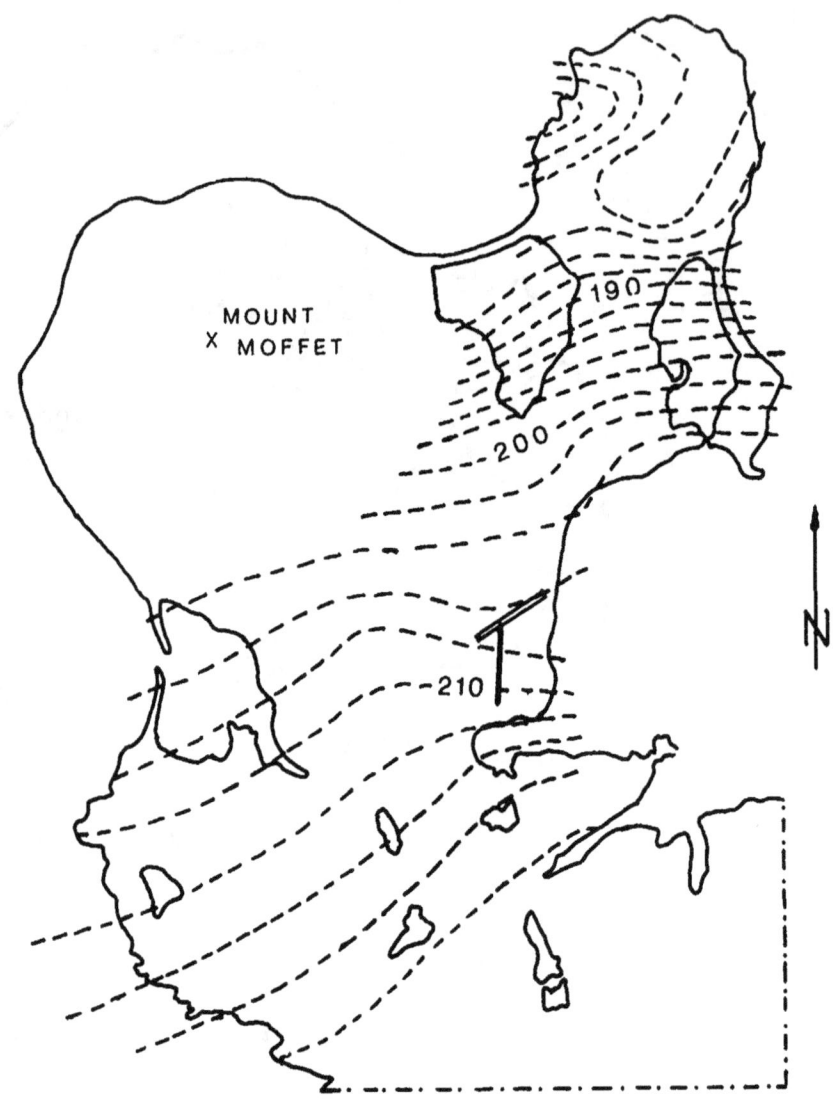

FIGURE 12. Bouguer Anomaly Map, Northern Adak Island, 1976.
Contour interval 2 mgals (modified from Hoover, 1976).

high are superimposed on the regional gradient. The USGS suggests that the gravity low in this area may represent a low-density zone associated with hot or perhaps molten rock.

The following is an excerpt from the summary as presented in the Hoover report (Reference 17):

"The gravity data indicates a large mass of low density rock below Adagdak. The gravity low has a northeast trend and is centered on the south to east side of Adagdak. The low density body evidenced in the gravity presumably is the same mass giving rise to the anomalous P delay (reported at a seismic station located near Alpine Lake on Mount Adagdak in comparison to other seismic stations on Adak — Personal communication to the USGS by R. Engdahl (1976)). The gravity also shows it is not at great depth although additional data is desired to better define the body.

"The MT data indicate again a large body and one with a northeast trend under Adak. The change in conductivity from the south part of the island to Adagdak is very large. If one assumes a bulk resistivity in the south of 500 to 1000 ohm-meters, a value probably too high, then a large zone of 1 to 2 ohm-meters underlies Adagdak. These values are quite low and definitely indicative of a geothermal source.

"The telluric survey at 20-30 second period defines the southern part of Adagdak as being anomalously conductive and defines a limited area to the northwest of the Communications Station where conductive rock is nearest to the surface. The 100 to 1 range in conductivity from the edge of Adagdak to the anomalous center zone also suggests very low resistivities in the area.

"The EM-16R and AMT surveys show the resistivity under Adagdak decreasing with depth. The lowest values being in the area defined by the telluric survey and that a very conductive zone is located there at about 600 meters in depth.

"The self potential data although quite speculative due to large noise voltages also can be interpreted as suggesting a large anomalous body below Adagdak at depths of about 2 to 3 kilometers.

"These geophysical data show that a low density, very conductive, mass of rock underlies Mt. Adagdak, that its top is quite near the surface, and that it is probably elongated in a northeast direction. The low values of resistivity also imply hot or molten rock. Additional work is definitely indicated to further identify this probable geothermal system."

OBSERVATION HOLE DRILLING

The Navy used the results of the Hoover report and previous studies to site two slim drill holes. Results of the drilling are as follows:

Two observation (stratigraphic) holes were drilled as shown in Figure 13. The drilling was done in the summer of 1977 by the Hamilton Drilling Company, which used a track-mounted Longyear drilling rig. Hamilton intended to core; however, the large amounts of clay alteration dictated rotary drilling. Both holes were completed with 2-inch capped black iron pipe cemented in and filled with water.

Hole 1 was sited by the USCS on the basis of its geophysical studies. It was drilled to 995 ft on the southwestern flanks of Mount Adagdak. Much to the surprise of the NWC and USGS personnel, the entire hole was drilled in volcanic ash altered to montmorillonite with occasional volcanic boulders or flows. A maximum recording thermometer lowered to the bottom of the hole on 12 January 1978, and left there for 2.25 hours, gave a reading of 75°F. Using a mean air temperature of 40°F (Reference 1, p. 376) would give an overall thermal gradient of 3.5°F/100 ft (6.4°C/100 m).

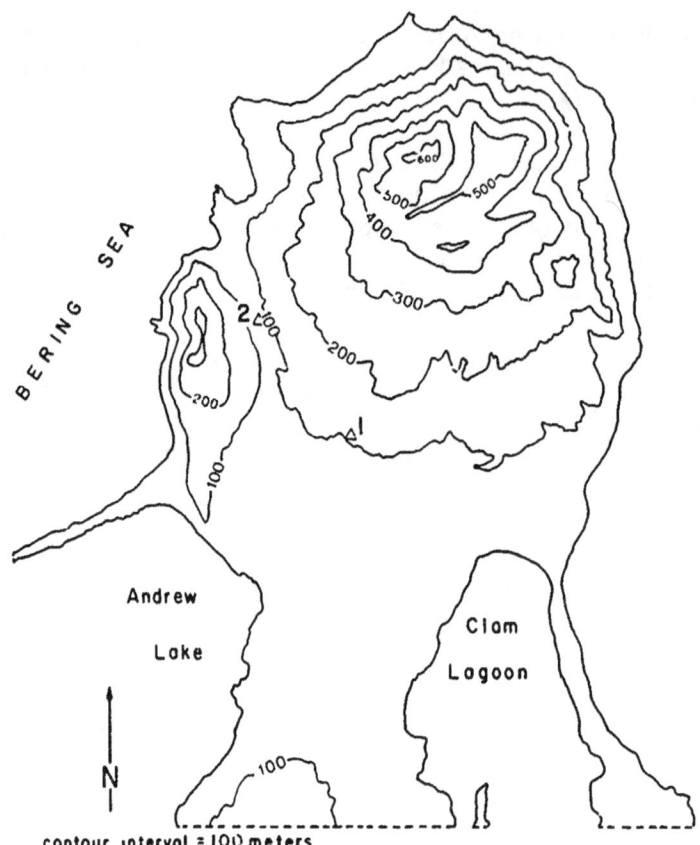

FIGURE 13. Locations of Observation Hole 1 and
Observation Hole 2.

The second hole was located by NWC by projecting the trend of the most prominent joint pattern at the Andrew Bay Hot Springs into the geochemical and geophysical anomalies. The hole was also located along a road for environmental reasons. It was drilled to a depth of 1,925 ft, through competent but fractured rock.

A temperature log on this hole is shown as Figure 14. Detailed data are contained in Appendix C. The gradients (from 50 to 1,925 ft) of 4.2°F/100 ft (7.6°C/100 m) are not spectacular but are encouraging. The gradient curve is a straight line, indicating conductive heat transfer. Extrapolating the curve would indicate that a temperature of 190°C would be attained at a depth of about 4,000 ft.

ADDITIONAL GEOPHYSICAL SURVEYS

In the summer of 1982 NWC conducted additional gravity and magnetic surveys; made a reconnaissance soil survey of mercury content; and contracted with the Earth Sciences Laboratory, University of Utah Research Institute, for a microseismic survey. All of the surveys were confined to the northern part of the island.

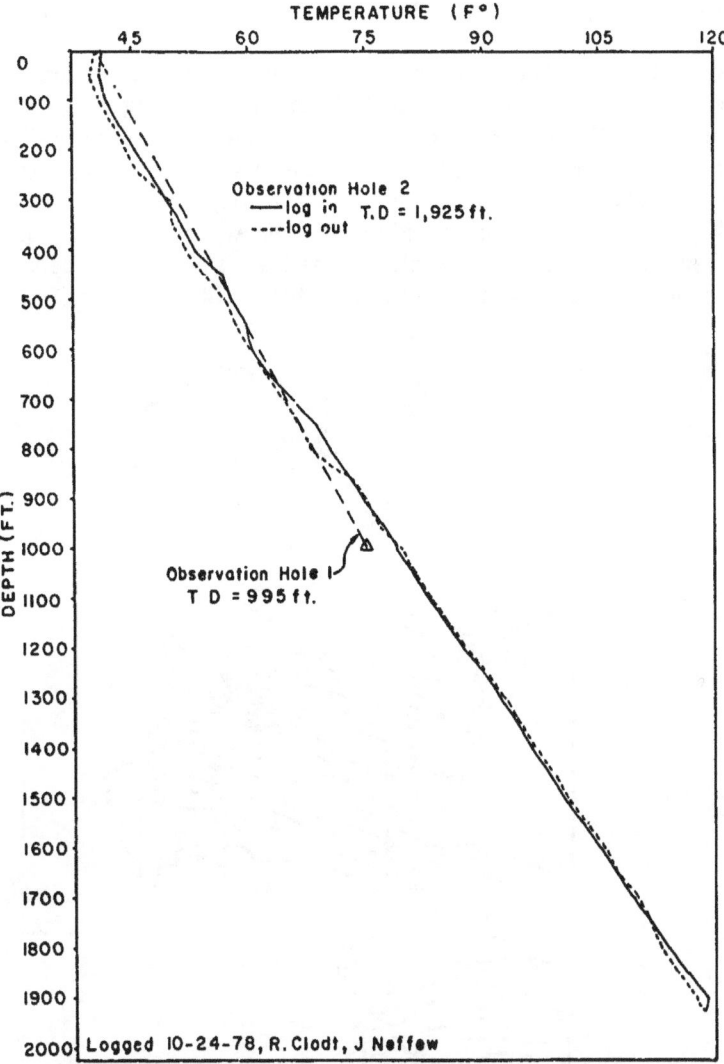

FIGURE 14. Graph of Temperature Log From Observation
Hole 1 and Observation Hole 2. Graph of Observation Hole
1 based on single point at total depth (T.D.)

Gravity II

A total of 301 gravity and land magnetic stations were occupied in July and August 1982. Fifty-three stations were taken at existing bench marks on the island, and 248 by-stations were created using a Wild T-1 theodolite and two wide-faced rods. Surveying of the new stations took place despite unfavorable weather of torrents of rain alternating with gale-force winds; elevation accuracies were better than 0.4 ft. Gravity was measured at each station by a LaCoste and Romberg gravity meter (Model G-144), in a series of 4-hour drifts with checkpoints, and tied to the Adak Base Station. The findings were then reduced with the post-1967 latitude correction (Reference 18), assuming a reduction density of 2.67 gm/cc. Terrain corrections were taken in the field to a distance of 175 ft (Zone C of a Hammer chart in Reference 19), manually with a map through about 8,600 ft (Zone H of a Hammer chart). and then with a computer through approximately 72,000 ft (Zone M).

Figure 15 shows the Navy's gravity survey, plotted using a reduction density of 2.67 gm/cc and contoured on an interval of 1 mgal. Comparison of Figure 15 with the gravity survey reported by Hoover (Figure 12) does not show any great dissimilarity. The northward decrease of the Bouguer anomaly values appears the same, although the large gravity gradient south of Mount Adagdak shows slight tightening. It is not known how far north the decrease extends, however, because the Navy's survey did not include any points on the flanks of Adagdak due to the lack of survey time and bad weather. The Navy's survey did delineate the localized high near the Andrew Bay Hot Springs as seen by the USGS.

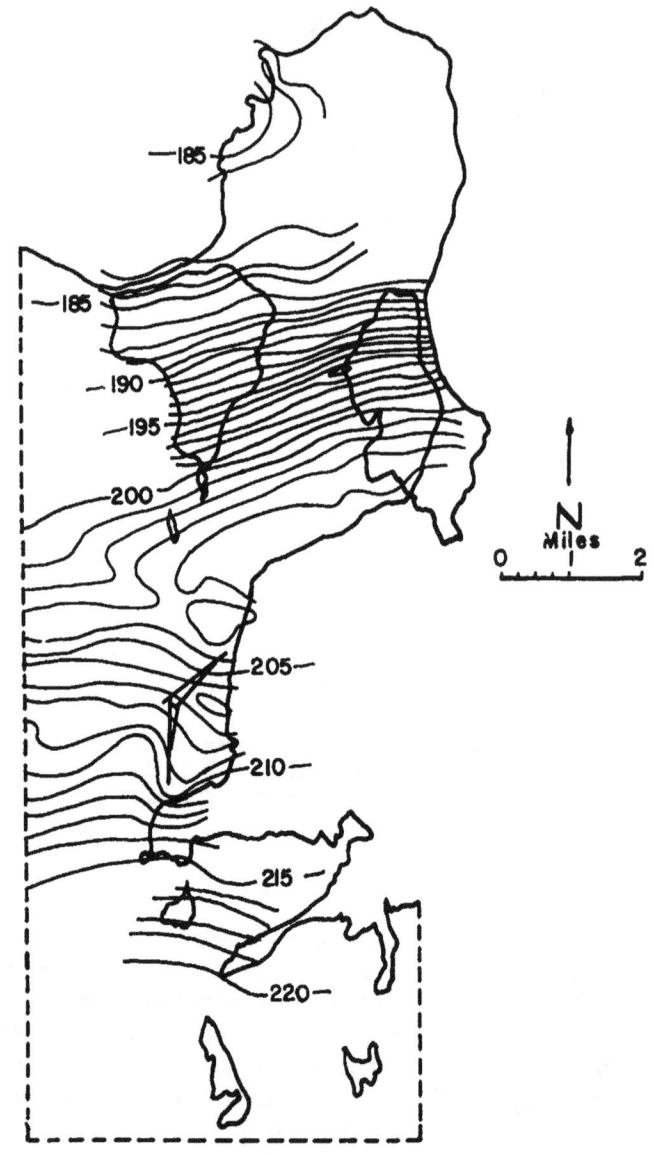

FIGURE 15. Bouguer Anomaly Map, Northern Adak Island, 1982. Contour interval, 1 mgal; reduction density, 2.67 gm/cc.

The Navy defined two localized lows in the vicinity of the Adak airfield. This finding corresponds well to the outcrops of andesite porphyry domes as mapped by Coats, and probably indicates that that rock type also exists at a shallow depth beneath the airfield (Reference 3).

A simple regional gravity trend was subtracted from the complete Bouguer gravity to reveal the residual gravity. The regional trend assumed a constant 2 mgal per mile decrease northward across the island. This regional gravity trend was discussed by Hoover and was based on information provided by Grow (References 17 and 20). A trend of N60°E was determined for the regional since this is the predominant structural trend on the islands of the Andreanof group (Reference 3).

The residual gravity map is shown in Figure 16. The predominant feature of the map is the northeast gravity trend that passes through both Andrew Lagoon and Clam Lagoon. Cross

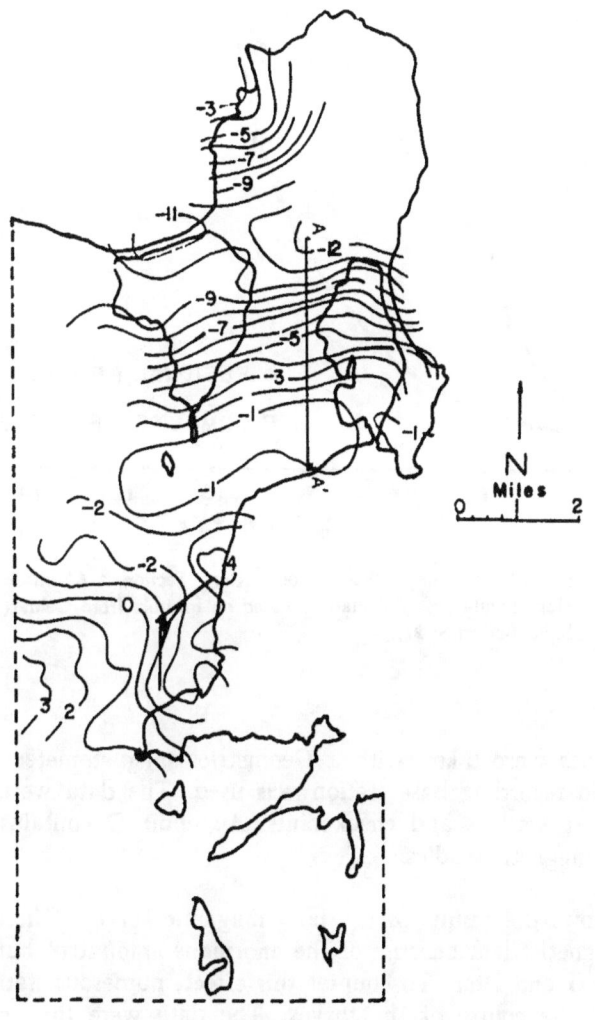

FIGURE 16. Residual Gravity Map of Northern Adak Island. Map constructed by subtracting from Figure 15 regional gravity gradient of 2 mgals/km trending N60°E.

section A-A' was drawn across the feature to attempt modeling. Using the fact that Observation Hole 1 drilled through almost 1,000 ft of clay, and that the two microseismic surveys delineated a fault structure dipping at nearly 70 degrees to the north, the feature was modeled as a fault assuming a density contrast of 0.4 gm/cc (2.2 gm/cc for the clay; 2.6 gm/cc for the Finger Bay volcanics). Results of the modeling are shown in Figure 17. This figure shows that the feature resembles a dip-slip fault with a throw of nearly 2,800 ft. The excess mass near the center of the fault model could represent a slide-block formed during faulting. The residual map of Figure 16 also clearly shows the gravity high found near the Andrew Bay Hot Springs.

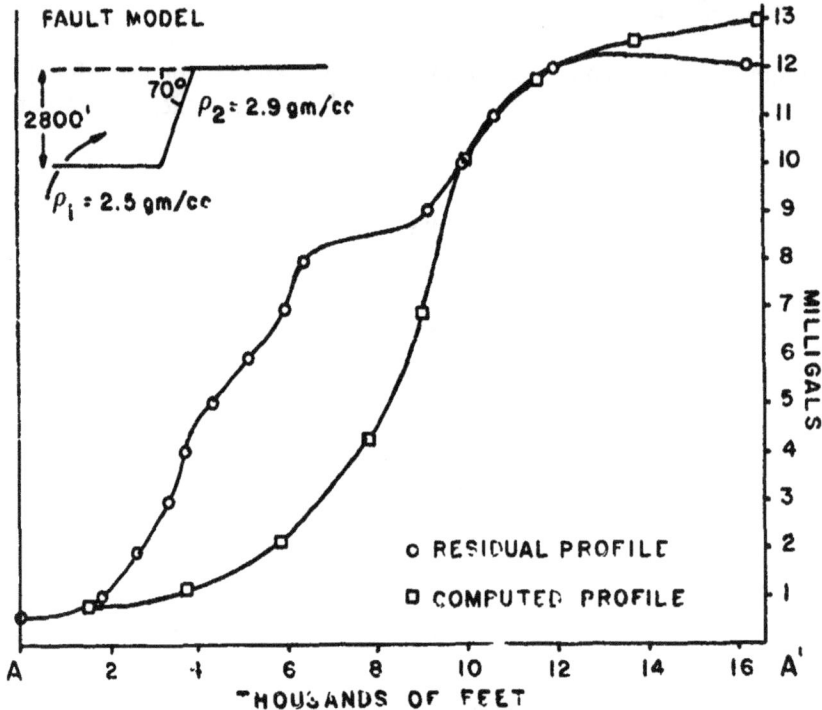

FIGURE 17. Simple Fault Model Along Section A-A' on Residual Map (Figure 16). Calculations based on formula from Geldart et al, 1966 (Reference 21).

Land Magnetics

Land magnetics were taken with a Geometrics magnetometer in conjunction with the gravity survey. No recording base station was used. The data were smoothed by repetitive recording from base stations and checkpoints. Appendix D contains all data on the Navy's gravity and land magnetic studies.

Figure 18 shows the results of the land magnetic survey. There was some difficulty in gathering the magnetic data because of the enormous amount of buried ferrous material left from World War II and later. To counter this effect, numerous stations were occupied more than once during the course of the survey. The data were then smoothed by using these checkpoints. Data that were obviously affected by possible buried material were deleted. The data were then plotted and contoured on an interval of 500 gammas.

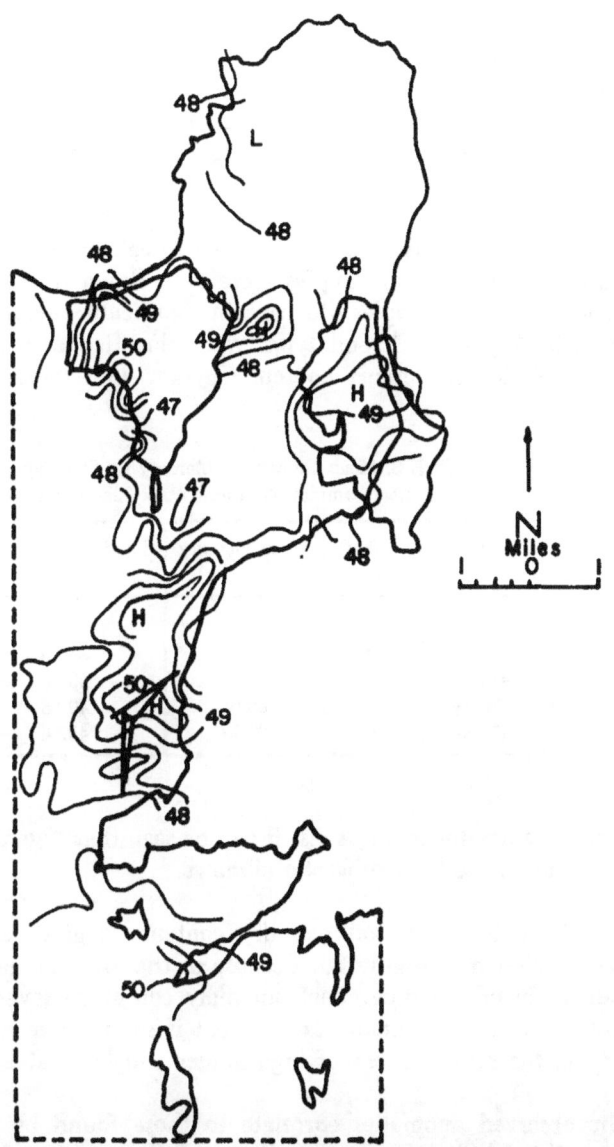

FIGURE 18. Ground Magnetic Map of Northern
Adak Island. Contour interval in gammas. All
values times 1,000. H = magnetic high anomaly;
L = magnetic low anomaly.

The most prominent features shown by the magnetic survey are the localized highs located near the airfield. These highs correlate well with the outcrops of andesite porphyry domes discussed in conjunction with the gravity survey, and they are probably the magnetic signature of the accessory magnetite that Coats describes within the groundmass of the porphyry (Reference 3). Another high exists at the site of the old airfield south of Mount Adagdak, but it is not certain whether this high is due to the existence of a buried andesitic dome or the magnetic expression of near surface metal artifacts. Other prominent features indicated are the faint outlines or suggestion of magnetic highs within Clam Lagoon and Andrew Bay, as well as an emerging low on the western flanks of Mount Adagdak.

31

SOIL GEOCHEMISTRY

Soil sampling was conducted to determine mercury content; mercury often occurs in unusually high amounts in soils over geothermal resources.* A reconnaissance soil survey was run on the northern end of Adak Island. The writers knew of no references to soil geochemistry in tundra and had reservations as to its effectiveness, as Hoover had reported high resistivities in the surface layer of soil on Adak because of excessive leaching (Reference 17). Samples were collected from the base of the surface vegetation, if present, or just below the surface if vegetation was not present. A plastic scoop was used to collect samples, which were placed in plastic zip-lock sandwich bags. The samples were then flown to NWC and analyzed with a Jerome mercury detector. The samples did contain significant amounts of mercury; data and locations are given in Appendix E. Soil types were classified as gravelly, sandy, ashy, and mucky (highly organic). Mean mercury content together with standard deviations is given in Table 3.

TABLE 3. Mean Content of Mercury by Soil Type, Geochemical Samples, Northern Adak Island, Alaska.

Soil type	No. samples	Mean mercury content, ppb[a]	Standard deviation, ppb
Gravelly	6	30.2	11.63
Sandy	16	41.0	34.40
Ashy	6	46.8	40.85
Clayey	35	69.0	40.98
Mucky	11	87.27	40.98

[a] ppb = parts per billion.

From the Table 3 entry for mucky soils, it can be seen that organic matter is an important factor in fixing mercury in soils in subarctic climates.

A histogram of frequency of soil mercury contents is given as Figure 19. Contoured mercury contents are shown as Figure 20. Because of the low number of samples, contouring was done independently of soil type. Each anomaly contained several soil types. Therefore, although the contours would be somewhat different if sample size had been large enough to contour by soil type, the same pattern of high values and low values would exist.

Many of the observed anomalies correlate to those found by the gravity and ground magnetic surveys. The low mercury anomalies at the airfield and at the communications station are probably the signature of buried or partially buried andesite porphyry domes. The anomaly at the mouth of Clam Lagoon directly west of Zeto Point is probably also a surface expression of a buried andesite porphyry dome, although it is not clearly shown by the geophysics. The high mercury anomalies directly west and south of the airfield lie in the highlands and are likely to represent volcanic rock of the Finger Bay formation. The high located on the western flanks of Mount Adagdak may also represent the interbedded lava flows as mapped by Coats and reported in Reference 3. However, the drilling that was done in this area indicated that over 1,000 ft of altered volcanic material is present. This signature then could be the surface expression of geothermal alteration of volcanic rock into clay. The migration of geothermal fluids tends to concentrate mercury directly above the fluid path.

* Mercury in soil is what is known as a "pathfinder element" for base metal deposits and epithermal (low-temperature) gold and silver deposits, as well as geothermal resources.

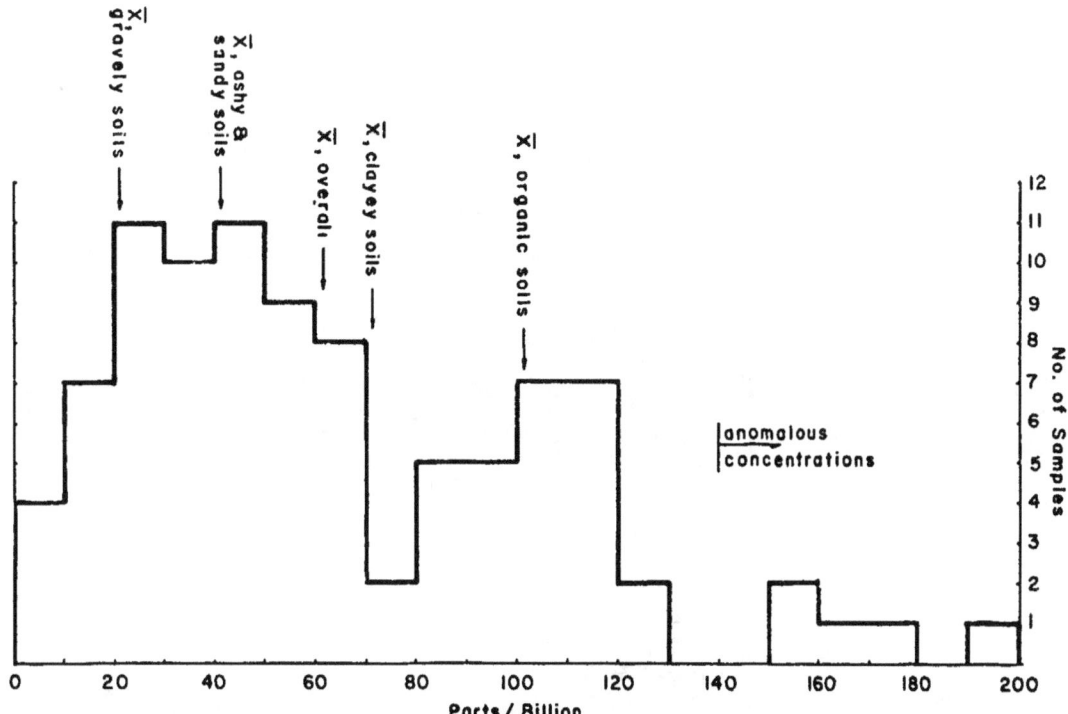

FIGURE 19. Frequency Distribution of Mercury Content in Soil Samples, Northern Adak Island.

A rock sample taken 200 ft above the Andrew Bay Hot Springs contained 283 ppb mercury. A rock sample from the spring itself gave a mercury content of 54 ppb.

Three samples were taken in the vicinity of the Kiguga Warm Springs. Sample 1, taken from just beneath the tundra, had a mercury content of 39 ppb, well within the background range. Sample 2, taken directly from the area of seepage at the silicified zone, had an extremely anomalous mercury content of 1,582 ppb. The third sample, taken from 30 yards to the south in an iron-stained silicified zone, had a mercury content of 316 ppb. Again, this is an anomalous sample. The type of alteration (sericitization and silicification, the introduction of pyrite, and the extent of alteration) makes the Kiguga Warm Springs area an attractive exploration target. The location, however, would present logistic problems both from the standpoint of exploration and utilization.

For selected samples, qualitative scans were run using a Kevex 0700 Energy Dispersive X-ray Spectrometer to check for heavy metals. If metals were present, the unit was used to check possible positive or negative correlations with the mercury content. Screened soil samples were placed in cups, covered with Mylar, then scanned. Because of interferences, only the following elements could be checked: silver plus indium, manganese, copper, and zinc. Counts per fixed time gave low correlation coefficients against mercury when fitted by least squares to linear, logarithmic, power, and exponential curves.

Surprisingly, no arsenic or antimony was found in any sample. These elements are generally indicative of shallow recent geothermal or epithermal fluid-migration zones.

33

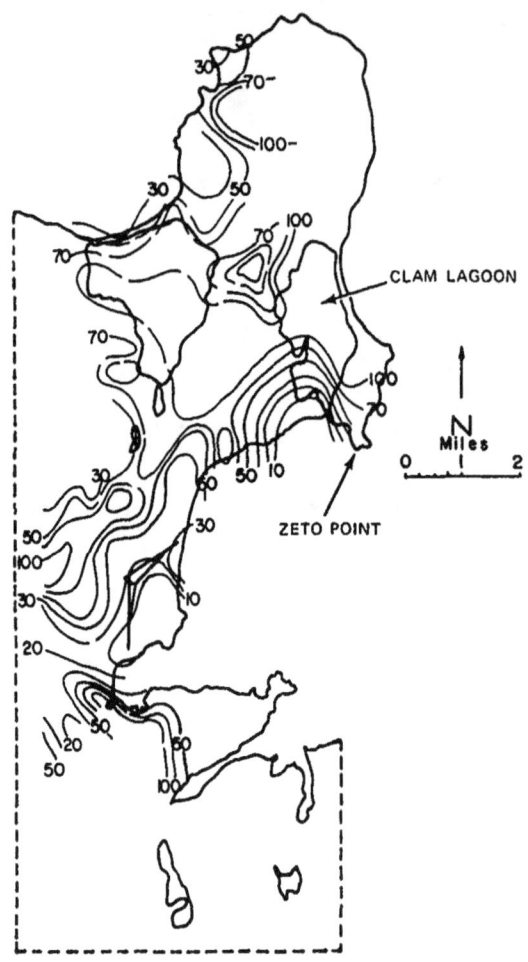

FIGURE 20. Contoured Mercury Concentrations in Soil Samples, Northern Adak Island. Contour interval in parts per billion.

MICROEARTHQUAKES II

During 1982 NWC contracted with the Earth Science Laboratory, University of Utah Research Institute to perform a microearthquake survey of northern Adak. The survey was meant to substantiate the seismicity survey of the same area by Butler and Keller (Reference 15). The survey, undertaken in early fall, recorded 190 discrete seismic events, of which 33 were determined to be of local origin. Of these seismic events, 24 were located. The events were analyzed when recorded on two or more stations of a nine-station network. Results were reported by Lange and Avramenko (Reference 22).

The hypocenters of the located events delineated a fault plane dipping north-northwest beneath Mount Adagdak toward the Bering Sea (Figure 21). The surface projection of the fault appeared to pass through Clam Lagoon. However, these determinations were made assuming a 5 km/s constant-velocity earth model and would probably change if a more realistic earth model were used. The survey does support the north-northwest-dipping fault plane deduced from the 1974 survey except for the fact that the trace lies north-n   .west of the one located in a later survey (References 15 and 22).

34

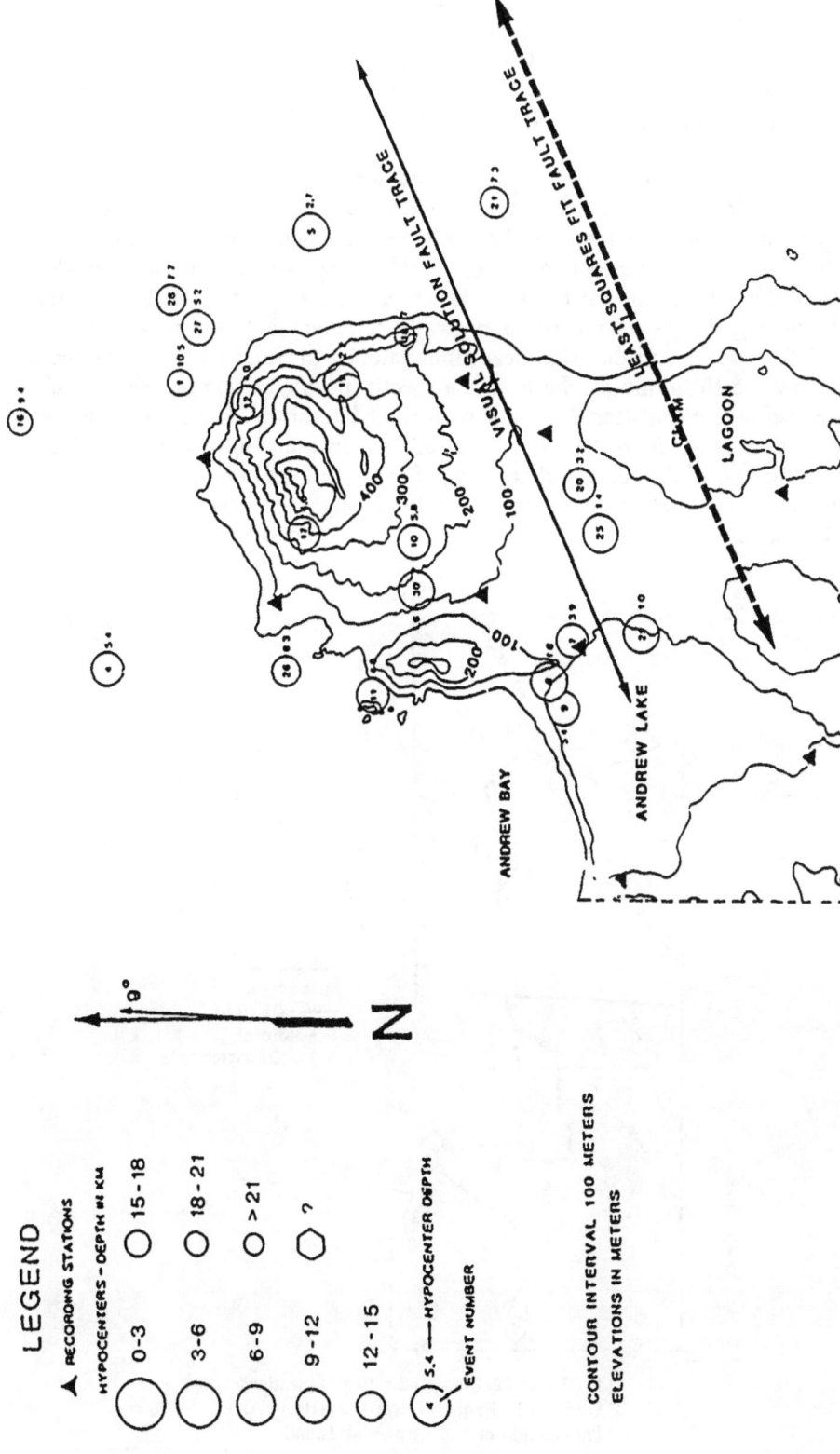

FIGURE 21. Results of Second Microseismic Study on Mount Adagdak. Model assumes constant seismic velocity of 5 km/sec. From Lange and Avramenko, 1982.

AERIAL PHOTOGRAPHY INTERPRETATION

A study of the northern part of Adak Island by means of high-quality aerial photography (scale 1:76000) was attempted to delineate any faults or lineations. While much of the structure appears to be obliterated by the presence of the volcanoes, some interesting features did present themselves. Results of the study are shown by Figure 22.

The most obvious feature on the aerial photographs is an east-west-trending fault that passes south of Mount Moffett and traverses the Naval Air Station just north of the airfield ("A", Figure 22). Since the trace of the fault cuts through flows from Mount Moffett and apparently shows some movement (probably south side up, strike movement not determinable) this may have been the fault that Professor Bruce Marsh was looking for during a study in 1981 of airfield runway cracks on the north runway (Reference 23). Marsh states that, during a study in 1973, it was found that significant inundation had occurred during earthquakes along the beach just north of the southern two andesitic porphyry domes (Figure 23). Although Marsh assumed a northeast-trending fault with right lateral offset to explain the separation of the southern two domes from the northern three, he concluded after a study in the area of the southern domes and of the runway that no significant fault could be found. However, the fault found on the aerial photographs passes directly through the area of inundation and could possibly be active.

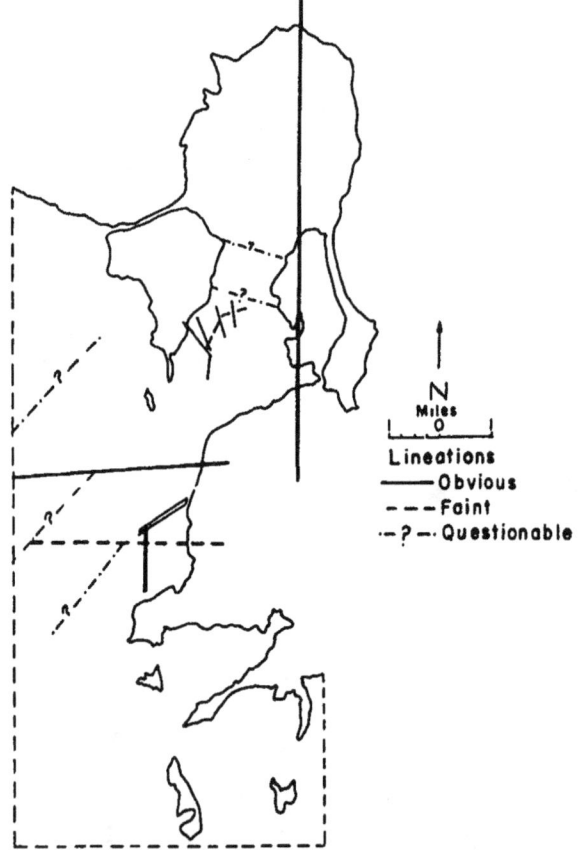

FIGURE 22. Map Indicating Lineations Deduced From High-Quality Air Photographs of Northern Adak Island.

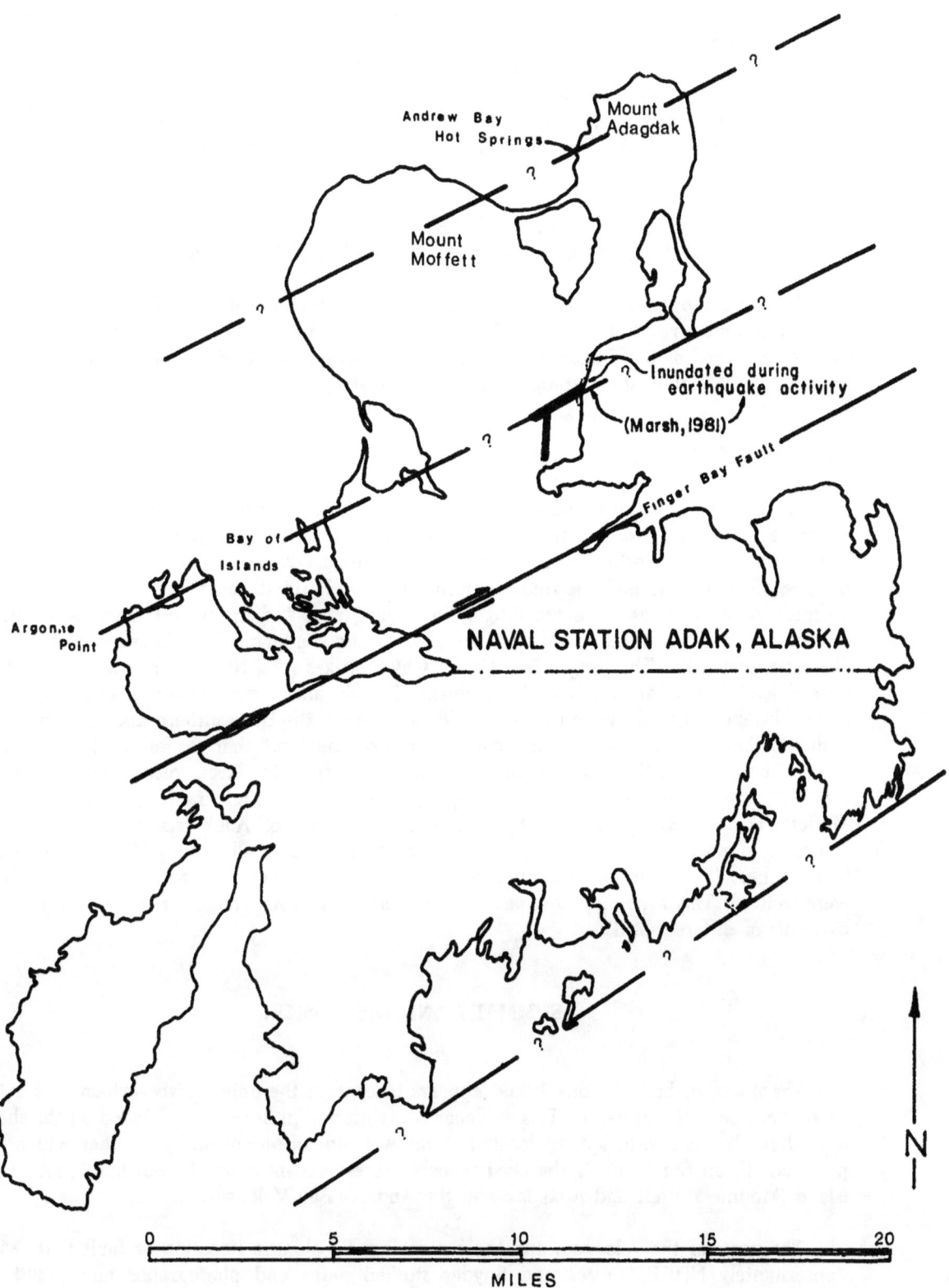

FIGURE 23. Map of Adak Island Indicating Possible en Echelon Faulting on a Strike of N60°E.

Another fault that was located by examination of the aerial photographs is a north-south-trending fault that cuts Mount Adagdak. It can be traced from Cape Adagdak south, passing directly east of the largest basalt/andesite dome on Mount Adagdak, through Clam Lagoon, where it probably is responsible for the existence of the small "island" within the lagoon, and then on out into Kuluk Bay. If the trace is extended far enough to the south, it passes through Thumb Bay just east of Finger Bay. The fault is probably slightly east side up, but any strike movement is unknown.

Other obvious faults as seen on the aerial photographs are those on the hill between Andrew and Clam Lagoons. These were mapped by Coats (1956, Reference 3) and others.

Figure 22 also indicates the locations of faint lineations indicated by the photos. Although cut by the faults previously mentioned, a great number of the lineations strike northeast and parallel the trend of the Finger Bay Fault. Other lineations strike due east-west while a third set near the communications station strikes northwest.

OTHER OBSERVATIONS

The Finger Bay Fault is one of the most pronounced structural features on the island. It is traceable on the large-scale topographic maps from Lucky Point west through the Bay of Islands and into the North Arm of Three Arm Bay (Figure 23). As a mental exercise, consider the possibility of "pushing" the islands within the Bay of Islands, along with the main mass of northern Adak, 3 miles along the Finger Bay Fault trace and up against Argonne Point. In essence, the Bay of Islands would close up, in a closing similar to that of oceans in the theory of continental drift. The Finger Bay Fault, which strikes in a N60°E direction, would then have at least 3 miles of right lateral movement. This distance, coincidentally, is the same right lateral distance when drawn on a line of N60°E between the two southern and three northern andesitic domes on Kuluk Bay and could represent the fault that Professor Marsh thought active. The direction N60°E also defines the line drawn from the large volcanic vent on Mount Adagdak, through the Andrew Bay Hot Springs, and then through the large vents on Mount Moffett. It appears possible that the entire northern half of Adak Island once moved, or continues to move, in a series of right lateral, en echelon faults, all striking N60°E (Figure 23). One of these faults may have defined a large zone of weakness, where the magmatic fluids were emitted. This zone may now emit geothermal fluids, even though it could be broken up by faults of different strike.

SUMMARY AND DISCUSSION

The northern half of Adak Island appears to contain the only surface evidence of active geothermal potential at depth. This evidence is manifested in hot springs located on the shores of Andrew Bay and warm springs located on the west side of Mount Moffett. Other evidence of past activity on the island is the occurrence of three volcanoes on the northern part of the island (Mounts Moffett and Adagdak and the Andrew Bay Volcano).

The vents of the volcanoes and the hot springs lie along a lineation or fault that strikes approximately N60°E. Fraser and Snyder studied maps and photographs and found the lineations of the southern half of Adak Island are strongest to the northeast (Reference 1).

Joints measured in southern Adak that strike northeast are vertical or dip steeply to the northwest. N60°E corresponds closely to the predominant direction that Coats found on other islands located within the Andreanof group lying to the east of Adak (Reference 3). This trend also coincides with the direction of the Finger Bay Fault, where there is some evidence of certain sections of the fault having a right lateral movement of nearly 3 miles. Indeed, Fraser and Snyder suggested that the Finger Bay Fault and the major Adak fault of Bradley were instrumental in blocking out or segmenting the islands (References 1 and 2 respectively).

Results of two microseismic studies 8 years apart indicated that a fault trending approximately N60°E and dipping more or less 70 degrees to the northwest was present, either cutting through Mount Adagdak or directly south of the mountain. Results of two gravity surveys tend to confirm this interpretation. Using information from a drill hole as a parameter, modeling of the gravity information indicated that the fault may be down-dropped to the north by as much as 2,800 ft. However, field work attempted by Marsh (1981) and Brophy (1982) did not uncover any surface evidence of the existence of the fault, nor did aerial-photograph analysis by a number of investigators (although the authors in their analysis did discern faults trending east-west and north-south in that area) (References 23 and 13).

The Andrew Bay Hot Springs lie on the southern flanks of an unclosed gravity high discerned by the two separate surveys. Directly east and south of the high lies a large gravity low. This low, coupled with the results of numerous electrical surveys, led the USGS to surmise that the subsurface of Mount Adagdak could be hot or perhaps molten at a shallow depth. Guided by the USGS theory and expecting a few feet of tundra and volcanic ash followed by competent rock, NWC drilled a slim hole at a site preselected by the USGS. To the surprise of everyone, the hole contained nearly 1,000 ft of altered volcanic ash in the form of clay. Since the drill hole lies in the trend of a fault inferred by numerous surveys, it seems possible that circulating hot geothermal fluids have altered all rocks within the fault plane into clays. This geothermal alteration would account for the low-density rock seen by the gravity surveys and the highly conductive rock measured by the electrical studies. The thermal gradient measured in the drill hole was 3.5°F/100 ft.

A second drill hole was located north of the first, approximately three-quarters of a mile east of the Andrew Bay Hot Springs and into the flanks of the gravity high. The hole was drilled to a depth of 1,925 ft, and competent but fractured rock was encountered throughout the entire depth. The thermal gradient of this hole measured 4.2°F/100 ft.

The Andrew Bay Hot Springs are almost 200 ft apart on the east side of Andrew Bay and have measured temperatures in excess of 158°F. Water samples taken from the springs are essentially brines, probably because of an oceanic water component. The total of dissolved solids in the Andrew Bay Hot Springs is about two-thirds that of oceanic brines. However, the springs do contain a high concentration of silica and boron along with high values of bicarbonate and calcium. Miller and Smith suggest that the high concentrations of bicarbonate and calcium are due to the leaching of volcanic rocks (Appendix A). The fact that the springs also have a high silica content may indicate that the gravity high detected near the hot springs may be due to silica deposition within the near-surface rock. If this is the case, the gravity high would be the next logical target to drill since it may represent the cap rock over the geothermal reservoir.

Supporting the above interpretation is the occurrence of an unclosed aeromagnetic low seen on the second derivative map in this area. If the gravity high does indicate the deposition of silica material by geothermal fluids, then this aeromagnetic low may indicate the pyritization of magnetite minerals by those same fluids. A magnetic low is also seen on the ground magnetic map in this area and may indicate the same phenomenon. The high mercury content of soils on the western flanks of Mount Adagdak are a fourth likely indicator of near-surface geothermal fluids.

Geothermometry results indicate reservoir temperatures of 155 to 186°C for the Andrew Bay Hot Springs. Motyka calculated a deep reservoir temperature of 160 to 188°C. Motyka, in a personal 1985 communication to the authors, states that the hydrothermal system is influenced by magma. He determined this by use of helium ratios and analyses of ^{13}C in fumarolic CO_2.

The Cape Kiguga Warm Springs are located on the beach, 1 mile south of Cape Kiguga. Measured temperature at these springs was 60°F, and they had a pH of 4.5 (Reference 4). Iron-mineral depository springs were seen along the margins of the main spring area and extended southerly along the beach for about 1 mile. Brophy also visited the warm springs and thought that, because of the degree and extent of alteration and silicification, the springs may have been a former major hot spring system (Reference 13). He thought that they had been larger, at one time, than the Andrew Bay Hot Springs currently are. Since no geophysical work was done in the vicinity of the Cape Kiguga Warm Springs, it is not known what type of geophysical signatures, if any, might characterize the area. Mixing models gives the Kiguga Warm Springs a deep reservoir temperature of 54 to 91°C.

Elsewhere on northern Adak Island, the andesitic porphyry domes are characterized by low gravity, low mercury, and high magnetic anomalies. These geophysical anomalies are located along Kuluk Bay, in good agreement with the domes mapped by Coats in Reference 3. This rock type probably underlies the Naval Air Station facilities more extensively than previously thought. The ground magnetics and mercury studies indicate that another dome may underlie the communications station. The gravity signature of this dome is apparently masked by the lower density subsurface rocks, as it is not delineated on the gravity surveys.

CONCLUSION

Proof of an economic geothermal resource on Adak Island can be verified only by further drilling and actual production. The geological, geophysical, and geochemical data clearly indicate that a geothermal prospect worthy of drilling exploration is present on the northern part of Adak. The data also indicate that the western flank of Mount Adagdak, near the shores of Andrew Bay, has the highest potential for developing a geothermal resource at modest depths. This site presents a favorable drilling target that should be tested by drilling to a depth of not less than 4,000 ft; a depth of 6,000 ft would be considered optimal.

REFERENCES

1. G. D. Fraser and G. L. Snyder. "Geology of Southern Adak Island and Kagalaska Island, Alaska," USGS Bull. 1028-M (1959), pp. 371-408.

2. C. C. Bradley. "Geologic Notes on Adak Island and the Aleutian Chain, Alaska," *Am. Jour. Sci.*, Vol. 246, No.4 (1948), pp. 214-40.

3. R. R. Coats. "Geology of Northern Adak Island, Alaska," USGS Bull. 1028-C (1956), pp. 45-67.

4. J. L. Moore. Adak Geological Investigations; draft of trip report. California Energy Company, Inc., Santa Rosa, Calif. 1981.

5. Robert O. Fournier. "Application of Water Geochemistry to Geothermal Exploration and Reservoir Engineering," in *Geothermal Systems: Principles and Case Histories*, ed. by L. Ryback and L. J. P. Muffler. Chichester, England, John Wiley & Sons Ltd., 1981. Pp. 109-43.

6. A. H. Truesdell. "Geochemical Techniques in Exploration," in *2nd Proceedings, U. N. Symposium on the Development and Uses of Geothermal Resources, San Francisco, 1975.* Vol. 1 (1976), pp. liii-lxii.

7. R. O. Fournier and A. H. Truesdell. "An Empirical Na-K-Ca Geothermometer for Natural Waters," *Geochimica et Cosmochimica Acta*, Vol. 37 (1973), pp. 1255-75.

8. R. O. Fournier and R. W. Potter II. "Magnesium Correction to the Na-K-Ca Chemical Geothermometer," *Geochimica et Cosmochimica Acta*, Vol. 39 (1979), pp. 1543-50.

9. T. Paces. "A Systematic Deviation From the Na-K-Ca Geothermometer Below 75 °C. and Above 10^{-4} atm. P_{CO_2}," *Geochimica et Cosmochimica Acta*, Vol. 38 (1975), pp. 541-44.

10. R. J. Motyka. "High-Temperature Hydrothermal Resources in the Aleutian Arc," in *Proceedings, Alaska Geological Society Symposium on Western Alaska Geology and Resource Potential*, 1982. Pp. 87-99.

11. A. H. Truesdell and R. O. Fournier. "Procedure for Estimating the Temperature of a Hot-Water Component in a Mixed Water by Using a Plot of Dissolved Silica Versus Enthalphy," *J. Res. U.S. Geol. Survey*, Vol. 5, No. 1 (January-February 1979), pp. 49-52.

12. R. O. Fournier. "Geochemical and Hydrologic Considerations and the Use of Enthalpy-Chloride Diagrams in the Prediction of Underground Conditions in Hot-Spring Systems," *J. Volcano. Geotherm. Res.*, Vol. 5 (1979), pp. 1-16.

13. J. G. Brophy. Summary of Geologic Field Work Conducted on Adak Island 7/28 to 8/6; informal report. Baltimore, Md., Johns Hopkins University, 1982.

14. F. Keller, Jr., J. L. Meuschke, and L. R. Alldredge. "Aeromagnetic Surveys in the Aleutian, Marshall, and Bermuda Islands," *Trans. Amer. Geophys. Union*, Vol. 35, No. 4f (1954), pp. 558-72.

15. D. L. Butler and G. V. Keller. Exploration of Adak Island, Alaska; Appendix B of *Geothermal Energy in the Pacific Region*. Grose & Keller Naval Research Contract N00014-71-A-0430-0004, 1975.

16. E. R. Engdahl. *Seismicity and Plate Subduction in the Central Aleutians*, Cooperative Institute for Research in Environmental Sciences, Univ. of Colorado/NOAA, 1975[?]

17. D. B. Hoover. Geophysical Techniques Applied to the Evaluation of the Geothermal Potential of Adak Island, Alaska; preliminary USGS report (unpublished), 1976.

18. W. M. Telford, L. P. Geldart, R. E. Sheriff, and D. A. Keys. *Applied Geophysics*, Cambridge, Cambridge University Press, 1976. P. 15.

19. S. Hammer. "Terrain Corrections for Gravimeter Stations,"*Geophysics*, Vol. 4 (1939), pp. 184-94.

20. J. A. Grow. "Crustal and Upper Mantle Structure of the Central Aleutian Arc," *Geol. Soc. Amer. Bull.*, Vol. 84 (1973), pp. 2169-92.

21. L. P. Geldart, D. E. Gill, and B. Sharma. "Gravity Anomalies of Two-Dimensional Faults," *Geophysics*, Vol. 31 (1966), pp. 372-99.

22. A. L. Lange and W. Avramenko. *Adak Island, Alaska Microearthquake Survey: Preliminary Hypocenter Determinations*, Subcontract for Earth Science Laboratory, University of Utah Research Institute, U.S. Dept. of Energy Contract No. DE-AC07-80ID12079, 1982.

23. B. D. Marsh. Field Work on Adak Island, Aleutian Islands, Alaska, July 29, 30, 31, 1981; informal report. Baltimore, Md., Johns Hopkins University, 1981.

Appendix A

GEOLOGICAL TECHNIQUES APPLIED TO THE EVALUATION OF THE GEOTHERMAL POTENTIAL OF ADAK ISLAND, ALASKA

by

Thomas P. Miller and Robert L. Smith
February 8, 1977

Introduction

The U.S. Navy contracted with the U.S. Geological Survey in early 1976 to make an evaluation of the geothermal resources available on the naval station on northern Adak Island in the central Aleutian Islands. The U.S.G.S. accordingly began work in the summer of 1976 with a two-phase program consisting of (1) a geologic study concentrating on the petrology, chemistry, and geochronology of the recent volcanoes in the immediate area and (2) a geophysical study. This part of the report deals with the results of the geological investigations, most of which were concentrated on Mount Adagdak because of the apparent youthfulness of the volcano, its stage of evolution, and its access. The field work was done by Dr. Thomas P. Miller and Robert L. Smith during the period July 26-August 9, the geochronology studies were done by Dr. G. Brent Dalrymple, and the chemical analyses were done under the direction of Brent Fabbi, all of the Geological Survey.

Regional Geologic Setting

Adak Island is located in the central Aleutian Islands and is part of the Aleutian arc, a seismically active volcanic arc-trench system that extends for more than 2000 km across the north Pacific. Over 60 major volcanic centers of Quaternary age occur along the northern edge of this arc and at least 40 of these have been active in the past 200 years (Coats, 1952). The volcanism appears to be the result of northward subduction of the Pacific plate beneath the North American plate with magma being generated along the Benioff seismic zone which lies some 100 to 125 km beneath Adak Island. The northern part of Adak Island contains three of these volcanic centers, Mt. Moffett, Andrew Bay, and Mt. Adagdak.

Summary

Information obtained from a study of the petrology, age and composition of the three volcanoes on the north side of Adak Island was used in an attempt to determine whether thermal anomalies may be present beneath the volcanoes. Of the three volcanoes, Mt. Adagdak appears to have the highest probability for the existence of an underlying thermal anomaly of significant size. Mt. Moffett has a lesser probability and Andrew Bay volcano has a relatively small probability for a thermal anomaly of any magnitude.

On the basis of geological studies alone, the geothermal potential of northern Adak Island can be considered as marginal only. The decision for further work, including drilling, should be based on an integration of the geophysical and the geological data and a cost analysis study of the benefits of relatively cheap geothermal energy versus the costs of further exploration.

Local Geology

Most of Adak Island is underlain by the Finger Bay Volcanics, a thick (at least 1500m) series of predominantly pyroclastic rocks and lava flows chiefly of basaltic and andesitic composition but including minor amounts of rhyodacite. The unit is moderately deformed, intensively altered, and was intruded by granodiorite plutons in Miocene time (about 13-14 m.y., Marlow and others, 1973). The unit is considered to be early Tertiary in age and part of the "early series" of exposed rocks found in the central and western Aleutians (Marlow and others, 1973).

The Finger Bay Volcanics are overlain by the Andrew Lake Formation, a sequence of marine sedimentary rocks more than 850 m thick and composed chiefly of tuffaceous sandstone, siltstone, and siliceous shale interbedded with mafic flows. The unit was considered by Coats (1956) to be Paleozoic in age but more recent work by Scholl and others (1970) has shown it to be of middle or late Eocene age. The Andrew Lake Formation appears to rest depositionally on the Finger Bay Volcanics and is best exposed along the eastern shore of Andrew Lake (fig. 1). Scholl and others (1970) suggest that the unit was probably laid down in a perched basin along the crest of an early Tertiary Aleutian ridge; possible analogues exist in the present ridge system.

Several small glaciated domes crop out in the vicinity of Kuluk Bay (i.e. Zeto Point). They are composed of hornblende andesite, intrude both the Andrew Lake Formation and the Finger Bay Volcanics, and have yielded K-Ar ages of about 5 m.y. (Marlow and others, 1973).

Volcanic and volcaniclastic rocks associated with Mount Moffett, Andrew Bay, and Mount Adagdak volcanoes unconformably overlie the Andrew Bay Formation. Mount Moffett and Mount Adagdak physiographically dominate the north part of Adak Island. Mount Moffett appears to be a typical andesitic stratocone some 10 km in diameter and composed of intercalated flows and tuffs with associated domes and parasitic cones. Andrew Bay and Mount Adagdak volcanoes on the other hand appear to be dominated more by large andesitic to dacitic domes emplaced in central vents of andesitic stratocones considerably smaller than Mount Moffett. All three volcanic centers have been assigned by Marlow and others (1973) to the "late series" of exposed rocks in the region and were considered to be of late Tertiary and Quaternary age. There are no known accounts of historic volcanic activity (i.e. <200 yrs.) associated with these three volcanoes.

Most of northern Adak Island was covered by Pleistocene glaciers which originated in the highlands in the southern part of the island and lapped well up on the flanks of Mount Moffett and Mount Adagdak; moraines from this glaciation have been found by Coats (1956) at elevations of 500 feet on Mount Adagdak. Valley glaciers were formed on Mount Moffett and appeared to head in cirques on the north and northeast side of the volcano.

A well-stratified layer of volcanic ash, locally as much as several feet thick, mantles much of the lower elevations of north Adak Island. The ash is post-glacial and thought to have come from volcanoes on nearby islands, particularly Kanaga and Great Sitkin. The presence of pumice-rich layers in the ash suggests they may have originated during the caldera-forming eruption of Kanaton caldera on Kanaga Island.

Hot springs have long been known to occur along the east side of Andrew Bay and were sampled during the course of this study.

Numerous steeply dipping normal faults of relatively small displacement occur south of the volcanoes. These faults generally strike N. 60° E. to N. 60° W. and from N. 20° E to N. 10° W. (Coats,

1956); they do not cut the Quaternary volcanoes. Many of the faults
shown by Coats on Mt. Adagdak appear to represent landslide scarps
on the unstable north flank of the volcano facing the ocean.

Geochronology

The three volcanoes on the north side of Adak Island were thought
by Coats (1956) to be late Tertiary to perhaps late Quaternary in age.
based on morphology and similarity to other volcanoes in the arc.
Cameron and Stone (1970) reported K-Ar age measurements of 0.0±0.16
m.y. and 0.0±0.23 m.y. for fresh olivine basalt collected from basal
members of Mount Adagdak volcano along the southeast sea coast. They
state that it is highly unlikely that the sample could be older than
500,000 yrs.

Marsh (1976) in a recent paper on Aleutian andesites reported a
K-Ar age measurement on a basal member of Andrew Bay volcano as
being less than 500,000 yrs but published no supporting analytical
data. Based on this date, Marsh considered Mount Moffett and its
parasitic cone as being perhaps no older than 150-200,000 yrs while
Mount Adagdak may be no older than 100,000 yrs.

Several new K-Ar age measurements were made on Adak rocks by
G. Brent Dalrymple as part of the present study and these are reported
in table 1. Three of the four age measurements on Mount Adagdak volcanic rocks appear to be of good quality and range from 140,000±
34,000 yrs to 342,000±17,000 yrs. The two andesite (76AMm207 and 211)
age measurements are consistent with each other but are both considerably younger than the date on the dacite dome which also appears to be

of good quality. The opposite situation is suggested by the field relations; it should be emphasized that with rocks as young as these, the upper limit of the K-Ar age dating technique is being approached. The limited number of age measurements together with the poor exposures leaves us unable to resolve this problem. A further discussion of this problem is contained in the petrology section. The age measurements do however, strongly suggest that the general range of volcanic activity at Adagdak occurred between 100,000 and 350,000 yrs ago.

Only one of the age measurements from Andrew Bay volcano is of good quality and that age (827,000±167,000 yrs) suggests that Andrew Bay volcano is indeed the oldest of the three volcanic centers. This sample was collected from a dome at the base of which hot springs occur (fig. 1).

The age sample from Mount Moffett is from one of the young andesite domes on the east side of the volcano; the age, although not of the best quality, suggests the dome is less than 250,000 yrs old. The main part of the volcano would be older, perhaps on the order of a few hundred thousand years. The K-Ar measurements suggest that volcanic activity at Moffett and Adagdak volcanoes overlapped and was in part contemporaneous, a suggestion first made by Coats (1956).

Petrology

Mount Adagdak is the remnant of a small andesitic stratocone which has gone through periods of vigorous dome-building activity in the central vent area. Only the south half of the original stratocone remains and the central domes are exposed and open to the sea along the north side of the volcano (fig. 1). Coats (1956) has suggested that the stratocone itself was built upon an older shield volcano composed chiefly of basaltic lava flows with a probable thickness of over 300 m. These flows rest on a foundation of flat-lying tuffaceous sandstone exposed in the sea cliffs on the east side of the volcano.

The lava flows in what Coats called the shield volcano are at least in part composed of olivine basalt. In addition to olivine they also contain augite, calcic plagioclase, glass, and accessory magnetite. Chemical analysis of samples from these flows are given in table 3 (Nos. 46AC36, 76AMm211, AD14) and confirm the basaltic character of the flows with SiO_2 being <50 percent.

The composite stratocone consists chiefly of tuffs with subordinate flows. They appear to be chiefly augite andesite in composition (Nos. 76AMm206, 46AC21, table 2).

The domes in the central vent area appear to represent at least three major dome-building eposides. They are light gray in color and composed of porphyritic hornblende-bearing andesitic dacite. The phenocrysts are calcic plagioclase and oxyhornblende with a groundmass of augite, plagiocalse, and altered glass(?). Chemical analyses are given in table 2 (Nos. 76AMm205, 204, 207, and 46AC175) and show a SiO_2 content of 61-62 percent. The domes are progressively younger towards the north and material from the dome building stage in the form of breadcrust bombs, debris flows, and mud flows mantle the western slope of the volcano.

The history of the volcano as presented above is somewhat at variance with the K-Ar measurements (table 2). According to the model presented here, the volcanic rocks exposed on the slopes of the volcano represent a pre-dome cone-building stage and should therefore be <u>older</u> than the domes in the central vent. The K-Ar ages however, indicate just the opposite.

Three possible explanations suggest themselves. 1) The sampled flows (one on the west side, the other on the southeast side) may represent younger events satellitic to the main volcano. Such a situation is not uncommon, but in this case is not supported by either our limited field studies or the mapping by Coats (1956).

2) The whole-rock K-Ar measurements are inexact. Although three of the four measurements appear to be of good analytical quality, they are close to the upper limit of the technique. The limited number of whole-rock age measurements is insufficient to resolve the problem; 3) The dome rocks contain excess radiogenic argon for unknown reasons and therefore give a spurious older age.

The geology and field relations do strongly indicate however, that the volcano did evolve far enough to form a high level magma chamber. The silicic dacitic domes were then emplaced from this high level chamber.

Coats (1952) description of Andrew Bay volcano suggests it was a typical andesitic stratocone composed of intercalated breccia, tuff-breccia, and lava flows. Much of at least the north end of the volcano appears from our brief examination to consist of remnants of andesitic domes and dome debris, including monolithologic breccia. Coats however, also mentions the occurrence of olivine-hypersthene basalt fragments as fragments in tuff-breccia high on the eastern side of the volcano. The material may represent part of the original andesitic-basaltic stratocone.

Most of the andesite domes consist of highly fractured greenish-gray, hornblende-augite andesite with a glass-rich groundmass. Chemical analysis of samples collected from near the north end of the volcano (Nos. 76AMm217A, 218, and AD72, table 2) show them to be typical andesites with 56-57 percent SiO_2.

Judging from the remnant of Andrew Bay volcano still existing, the volcano did not evolve or differentiate to the stage that nearby Mount Adagdak volcano did. No dacitic rocks were reported by Coats nor did we find any such material.

Mount Moffett volcano is a large composite stratocone measuring about 11 km in diameter and rising to an altitude of 1196 m. It is a typical andesitic volcano composed of intercalated flows and pyroclastic rocks with a parasitic cone of basalt and several andesitic domes on the eastern and southern flanks of the volcano. Because of a combination of poor weather and a lack of helicopter support, relatively little work was done on Mount Moffett during this study and most of the description of the volcano is taken from Coats (1956).

The composite cone consists chiefly of tuff and tuff-breccia with angular fragments as much as 1.5 m across; a few lava flows occur near the top. The pyroclastic rocks are composed of hornblende andesite and porphyritic basalt. Tuff-breccia beds are as much as 120 m thick and are locally interbedded with marine boulder conglomerates.

The lower and earlier lava flows are olivine basalt in composition (No. 46AC48, table 2); later flows consist of hornblende andesite and olivine-hypersthene basalt (Nos. AD44, 46AC275, AD49, table 2).

The domes on the flanks of the volcano are hornblende andesite (Nos. 46AC174, 76AMm222, table 2) and olivine-bearing hypersthene andesite or basalt. The parasitic cone on the northeast flank of Moffett is composed of olivine basalt flows, 1.5-3 m thick interbedded with coarse basaltic lapilli-tuff (Nos. 46AC158, AD36, AD39, table 2). Gabbro and basaltic agglomerate fill the vent of the cone.

Remnants of a tuff-breccia cone of hornblende hypersthene andesite rests unconformably on the lava flows of the Moffett cone near the summit and represents one of the last stages of activity of the volcano.

Hot springs

Waring (1917) mentioned the occurrence of hot springs on Adak Island in his compilation of hot springs in Alaska and Fraser and Snyder (1959) report hot springs occurring on the east side of Andrew Bay. During the present study, hot springs were noted at 2 principal localities about 60 m apart on the east side of Andrew Bay about 2 km north of Andrew Lake.

The host rock at both localities appears to be strongly fractured and altered andesite dome rock. Numerous oxidized and altered zones along the fractures attest to a long period of hot spring activity. The southernmost spring issues forth at sea level (best observed at low tide) through beach cobbles near the foot of a small (15 m in diameter) rock "knob" which is about 7 m from the sea cliffs. The orifice of the spring is completely covered by the beach cobbles and there are no surface signs of a hot spring (i.e., algae, iron oxide, etc.). Hot water bubbles up through the cobbles and is dispersed through them into the ocean. Only an occasional trace of vapor calls attention to the presence of a hot spring. A water temperature of $71^{\circ}C$ was measured in the warmest spot in the cobbles; no estimate of flow could be made.

A small pool, about 1 m by 0.5 m and 45 cm deep, occurs in a crevice of a sea cliff bench about 10 m away. This small pool has a distinctive red scum of Fe_2O_3 and a temperature of $35^{\circ}C$ measured in the mud at the base of the pool attests to its thermal nature.

The other principal hot spring area lies about 60 m north of the first spring and they are separated by ragged point of rock sticking out into the ocean. The springs issue out of tight fractures in a low flat bench carved out of the sea cliff and only a meter or so above low tide. The springs occur over an area of about 15 m by 7 m and issue from steeply dipping (60° S.) fractures striking N. $50-90^{\circ}$ W. A hissing sound is associated with many of the hot springs and fissures. Temperatures of $59-63^{\circ}C$ were measured and red coatings of Fe_2O_3 are again abundant. Some attempt has been made to make a small pool by damming up one of the springs with cement.

Water samples were taken from both spring areas for chemical analysis which are given in table 3. The waters are essentially brines which is not surprising in view of their proximity to the ocean. The high concentrations of silica and boron however, are much higher than that found in sea water and are more typical of hot springs associated with active volcanic belts. Bicarbonate and calcium are also high and suggest leaching of volcanic rocks.

Water chemistry has proved valuable in estimating subsurface temperatures, and the various techniques and approaches are described by Mahon (1970), Fournier and Rowe (1966), White (1970), and Fournier and Truesdell (1973). The most quantitative temperature indicators have been shown to be (1) the variation in solubility of quartz as a function of temperature and (2) the temperature dependence of base exchange or partitioning of alkalies between solutions and solid phases with a correction applied for the calcium content of the water (the Na-K-Ca geothermometer). There is some ambiguity and uncertainty in both methods, and in any particular region, subsurface information may be necessary to calibrate adequately, or choose between, the methods.

Lacking knowledge of subsurface reactions, we have calculated subsurface temperatures using both the quartz solubility (assuming conductive cooling) and Na-K-Ca geothermometers. For the quartz solubility geothermometer (Fournier and Rowe, 1966), the equation is (Fournier, oral commun., 1973):

$$-\log_{10} C_{SiO_{2(aq)}} = (1.309 \times 10^3/T) - 5.19,$$

where T = temperature in kelvins, and

C_{SiO_2} = concentration of silica in milligrams per litre.

For calculations of subsurface temperatures from Na-K-Ca concentrations (from Fournier and Truesdell, 1973), the equation for temperatures above 100°C is:

$$\log_{10}(m_{Na+}/m_{K+}) + 1/3\log_{10}(m_{Ca+}/m_{Na+}) = 1647/T - 2.240$$

where   T = temperature in kelvins, and

      m_{Na+} = molality of sodium ion,

      m_{K+} = molality of potassium ion, and

      m_{Ca+} = molality of calcium ion.

For temperatures below 100°C the equation is:

$$\log_{10}(m_{Na+}/m_{K+}) 4/3\log_{10}(m_{Ca+}/m_{Na+}) = 1647/T - 2.240.$$

The results of these calculations for the Adak hot springs are as follows:

Geothermometer

Sample no.	Quartz	Na-K-Ca
76AMm220	186°C	187°C
76AMm221	175°C (using 185 ppm SiO_2)	182°C
	143°C (using 110 ppm SiO_2)	

The indicated subsurface temperatures are relatively high for Alaskan hot springs (Miller and Barnes, 1976) but similar to those reported for some other Aleutian Islands hot springs. The temperatures are about at the minimum temperature (180°C) thought necessary to drive large steam-turbine generators (Muffler, 1973).

Geothermal Potential

We have made an estimate of the nature of the possible thermal anomalies beneath Mt. Adagdak and Mt. Moffett using the "Smith-Shaw" model (Smith and Shaw, 1975). This model uses an estimate of the

probable volume of high level magma chambers and the age and composition of the most recent eruption from these chambers to determine the
probable solidification state of each volcanic system and its relation
to a 300°C isotherm.

The application of the Smith-Shaw model is illustrated graphically
for Mount Adagdak volcano (fig. 3) and Mount Moffett volcano (fig. 4).
The small volume, old age, and relatively primitive composition of
Andrew Bay volcano indicates any underlying shallow magma chamber
has long since crystallized and cooled; the hot springs along the
margin of the volcano are more likely relative to a thermal anomaly
under Adagdak.

The pairs of lines in figures 3 and 4 are drawn to represent a
spectrum of cooling models that identify igneous systems that are now
approaching ambient temperatures (systems above lines 5 or 6), systems
that may now be approaching the post-magmatic stage (systems between
lines 3 and 4), and systems that probably still have magma chambers
with a large molten fraction (points below lines 1 or 2). The pairs
of lines represent the effect of shapes ranging from slablike to
equant for different cooling models. For all of these cooling models
the depth to the top of the magma chamber is assumed to be 4 km.

The Adagdak volcanic system depicted in fig. 3 is based on a
range in volume and radiometric age as determined in the present
study. The field covered by probably limits of error of volume and
age just intersects line 6 which is an estimate of the time required
before the central temperature of the solidified magma chamber has

fallen to 300°C. The plot shown in fig. 3 is based in part on the radiometric ages. If, as suggested by stratigraphic relations, the dacite cone is actually younger than the andesite flows, then the thermal anomaly may be of higher temperature than shown.

On the basis of these geologic studies, therefore, a thermal anomaly of some sort very probably underlies all or part of Mount Adagdak. The temperature of this anomaly, however, may be relatively low (i.e., much less than 300°C). The geochemical thermometry of the hot spring waters is in good agreement with a temperature of less than 300°C as they suggest temperatures of 175-180°C. The geothermal potential of Mount Adagdak volcano, based on geologic studies alone, can at the present time be considered only as marginal in terms of electrical energy; the possibility of energy available for space heating, however, would be higher.

The situation at Mount Moffett is somewhat different than at Mount Adagdak in that Moffett is a large andesitic stratocone with no known silicic vents. The plot shown in fugure 4 is a tentative first approach to the evaluation of andesitic cones that have no known silicic vents but which may be evolving toward a high level system. In other words, if such volcanoes have a high level chamber, what are the restrictions on chamber size? The maximum size (area) is considered to be a function of cone diameter. In this case an approximate size is estimated by analogy to Kanaton cone and caldera (area $= 25$ km^2) on adjacent Kanaga Island. The dome distribution on Moffett suggests (if these domes were dacite instead of andesite) a possible area of about 32.4 km^2. We therefore assume a probable maximum size of 30 km^2.

If the age of the most recent volcanic activity on Mount Moffett is approximately correct, then a thermal anomaly beneath Moffett is possible although the original magma chamber has probably crystallized. It should be emphasized that we have only a single K-Ar age determination from Mount Moffett and are tentatively assuming that Mount Moffett was, or is, evolving towards a high level system.

References Cited

Cameron, C. P., and Stone, D. B., 1970, Outline geology of the Aleutian Islands with paleomagnetic data from Shemya and Adak Islands: Univ. of Alaska Geophys. Inst. Research Rept., UAGR-213, 153 p.

Coats, R. R., 1952, Magmatic differentiation in Tertiary and Quaternary volcanic rocks from Adak and Kanaga Islands, Aleutian Islands, Alaska: Geol. Soc. America Bull., v. 63, p. 485-514.

Coats, R. R., 1956, Geology of northern Adak Island, Alaska: U.S. Geol Survey Bull., 1028-C, p. 47-66.

Fournier, R. O., and Rowe, J. J., 1966, Estimation of underground temperatures from the silica content of water from hot springs and steam wells: Am. Jour. Sci., v. 264, p. 685-697.

Fournier, R. O., and Truesdell, A. H., 1970, Chemical indicators of sub-surface temperature applied to hot spring waters or Yellowstone National Park, Wyoming U.S.A., in Proceedings U.N. Symposium on the Development and Utilization of Geothermal Resources, Pisa, 1970, Vol. 2, Part 1: Geothermics, Spec. Issue 2, p. 529-535.

Fraser, G. D., and Snyder, G. L., 1959, Geology of southern Adak Island and Kagalaska Island, Alaska: U.S. Geol. Survey Bull., 1028-M, p. 371-408.

Mahon, W. A. J., 1970, Chemistry in the exploration and exploitation of hydrothermal systems, in Proc. U.N. Symposium on the Development and Utilazation of Geothermal Resources, Pisa, 1970, Vol. 2, Part 2: Geothermics, Spec. Issue 2.

Marlow, M. W., Scholl, D. W., Buffington, E. C., and Alpha, T. R., 1973, Tectonic history of the central Aleutian Arc: Geol. Soc. America Bull., v. 84, p. 1555-1574.

Marsh, B. D., 1976, Some Aleutian andesites: Their nature and source: Jour. of Geology, v. 84, p. 27-45.

Miller, T. P., and Barnes, Ivan, 1976, Potential for geothermal-energy development in Alaska - summary, in Halbouty, M. T., Maher, J. C., and Lian, H. M., eds., Circum-Pacific Energy and Mineral Resources: Am. Assoc. Petroleum Geologists Mem. 25, p. 149-153.

Muffler, L. J. P., 1973, Geothermal resources, in Brobst, D. A., and Pratt, W. P., eds., Potential mineral resources: A geologic perspective: U.S. Geol. Survey Prof. Paper 820.

Scholl. D. W., Greene, H. G., and Marlow, M. S., 1970, Eocene age of the Adak Island 'Paleozoic(?)' rocks, Aleutian Islands, Alaska: Geol. Soc. America Bull., v. 81, p. 3583-3592.

Smith, R. L., and Shaw, H. R., 1975, Igneous-related geothermal systems, in White, D. E., and Williams, D. L., eds., Assessment of geothermal resources of the United States--1975, p. 58, 83.

Waring, G. A., 1917, Mineral springs of Alaska: U.S. Geol. Survey Water-supply Paper 418, 114 p.

White, D. E., 1970, Geochemistry applied to the discovery, evaluation, and exploitation of geothermal resources; Rapporteur's report: U.N. Symposium on Devlopment and Utilization of Geothermal Resources, Pisa, Italy.

Table 1.— Analytical data for potassium-argon age measurements for volcanic rocks from Adak Island, Alaska.

Sample no.	Material dated	K_2O† (wgt. %)	Weight (gms)	Argon $^{40}Ar_{rad}$ (mol/gm)	$100 \frac{^{40}Ar_{rad}}{^{40}Ar_{total}}$	Calculated age* (10^6 years)
Mount Adagdak						
76 AHm-203	Dacite	1.469 ± 0.005 (4)	10.809 / 8.802	nil / nil	nil / nil	< 1.0
76 AHm-204	Dacite	1.478 ± 0.011 (4)	20.530 / 20.719	6.78×10^{-13} / 7.49	10.5 / 16.6	0.342 ± 0.017
76 AHm-206	Andesite	0.835 ± 0.003 (4)	20.132 / 20.926	3.59 / 2.16	4.0 / 5.8	0.194 ± 0.039
76 AHm-211	Andesite	1.017 ± 0.002 (4)	20.011 / 20.652	2.18 / 1.98	2.7 / 3.2	0.140 ± 0.034
Andrew Bay						
76 AHm-210	Dacite	1.019 ± 0.004 (4)	18.002 / 10.244	13.56 / 10.55	4.0 / 2.9	0.827 ± 0.167
76 AHm-218	Dacite	1.124 ± 0.002 (4)	20.683 / 18.750	-- / 12.93	<1 / 1.1	0.796 ± 0.712
Mount Moffett						
76 AHm-222	Andesite	1.438 ± 0.003 (4)	19.813 / 20.686	-- / 2.94	<1 / 1.2	0.142 ± 0.121

†Errors are calculated standard deviations. Number of measurements are in parentheses.

*$\lambda_\epsilon = 0.572 \times 10^{-10}$ yr^{-1}, $\lambda_\beta = 4.963 \times 10^{-10}$ yr^{-1}, $^{40}K/K_{total} = 1.167 \times 10^{-4}$ mol/mol. Calculated ages are weighted means where weighting is by the inverse of the variance. Errors are estimated standard deviations.

Table 2 .-- Chemical analysis of hot spring waters, east side of Andrew Bay; analysis in parts per million (ppm).

	76AMm220	76AMm221
SiO_2	218	110 (215 w/o dilution)
Ca	1500	1300
Mg	70	110
Na	6800	6100
K	460	380
*HCO_3	420	430
SO_4	120	330
Cl	13500	12000
F	0.49	0.55
B	87	70
pH	7.4	7.5
Temp. (meas.)	63°C	71°C

* Total alkalinity as HCO_3.

Table 3.-- Chemical analysis of volcanic rocks from northern Adak Island, Alaska.*

| | Mount Adagdak volcano | | | | | | | | | | | Andrew Bay volcano | | |
	46AC 36	76AHm 211	AD 14	46AC 21	76AHm 206	76AHm 208	46AC 13	76AHm 205	46AC 175	76AHm 204	76AHm 207	76AHm 217A	76AHm 218	AD 72
SiO_2	48.03	49.9	49.93	52.95	55.8	59.5	59.61	61.4	61.46	61.8	61.9	56.6	57.4	57.63
Al_2O_3	17.90	19.2	19.46	19.03	18.9	17.2	16.89	17.0	17.35	17.0	17.0	16.1	17.0	16.13
Fe_2O_3	4.48	3.3	3.24	8.18	3.6	3.3	6.29	3.0	3.72	3.0	3.0	3.2	3.7	3.28
FeO	6.30	6.5	6.30	0.87	3.6	2.8	0.84	2 6	2.64	2.6	2.7	3.1	1.9	3.64
MgO	6.13	4.7	5.07	3.66	2.6	2.7	2.96	2.5	2.16	2.4	2.3	4.4	2.4	5.52
CaO	11.83	9.8	10.43	10.01	7.8	6.8	7.73	6.5	6.73	6.3	6.3	7.1	7.3	8.88
Na_2O	2.98	3.02	3.31	3.35	3.48	3.54	3.61	3.56	3.88	3.70	3.73	3.23	3.31	2.93
K_2O	1.00	1.03	1.05	0.92	1.01	1.33	1.15	1.39	1.53	1.44	1.42	1.28	1.34	1.22
H_2O^+	.04	.12	.09	.10	0.69	.35	.07	.26	0.19	<.01	.08	.68	.75	.16
H_2O^-	.10	.15	.09	.10	.29	.12	.19	.28	0.04	.09	.05	1.77	1.20	.22
TiO_2	.92	.92	.90	.56	.56	.58	.48	.50	.38	.49	.51	.51	.50	.54
P_2O_5	.30	.17	.23	.31	.21	.12	.23	.13	.32	.14	.14	.14	.17	.15
MnO	.17	.20	.18	.18	.21	.15	.14	.14	.16	.15	.15	.12	.13	.15
CO_2	--	.01	<.05	--	.03	.08	--	.03	--	.02	.05	.39	1.3	.08
Total	100.18	99	100.16	100.22	99	99	100.19	99	100.56	99	99	99	98	100.53

Mount Moffett volcano

76AMm 210	46AC 158	46AC 48	AD 36	AD 39	46AC 174	AD 44	46AC 275	AD 49	76AMm 222
59.3	50.16	50.81	50.87	51.39	54.62	55.27	55.57	56.34	57.8
17.3	19.98	18.69	16.12	18.87	17.04	18.12	17.77	17.17	17.4
2.7	3.89	4.03	4.76	4.96	3.47	3.90	3.86	4.83	3.4
2.6	5.70	4.76	4.54	4.83	4.76	4.11	4.76	2.76	2.9
2.4	4.21	4.19	6.83	4.38	3.90	3.73	3.81	4.80	3.0
7.0	9.84	10.62	10.84	10.16	9.53	8.53	7.64	8.97	7.3
3.30	3.68	3.57	4.16	3.09	3.55	3.68	3.61	1.32	3.65
1.24	1.46	1.32	1.34	1.31	1.35	1.48	1.67	1.89	1.33
.90	0.15	0.00	.04	.04	.13	.25	.25	.08	.05
.71	.08	.02	.00	.03	.11	.15	.30	.02	.05
.46	.68	.92	.75	.84	.84	.70	.68	.72	.57
.15	.43	.24	.19	.23	.26	.20	.34	.24	.16
.14	.18	.28	.16	.18	.15	.16	.18	.14	.15
1.2	--	--	<.05	.05	--	<.05	--	<.05	.05
99	100.44	99.45	100.6	100.27	99.71	100.28	100.44	99.28	98

Table 3 (cont.) Description of analyzed samples.

Adagdak

46AC36   - Basalt flow 0.7 mile northeast of northeast corner of Clam
            Lagoon (Coats, 1952).
76AMm211   - Basalt flow, near roadcut on northwest side of Mt. Adagdak.
AD-14   - Upper andesite flow exposed in roadcut on southwest side of
            Mt. Adagdak (Marsh, 1976).
46AC21   - Andesite dome 1.2 miles southeast of summit of Mt. Adagdak
            (Coats, 1952).
76AMm206   - Andesite thick massive flow, sea-cliff on southeast side
            of Mt. Adagdak.
76AMm208   - Andesite dome, northeast side of Mt. Adagdak.
46AC13   - Andesite lowest flow, sea cliff on southeast side of
            Mt. Adagdak (Coats, 1952).
76AMm205   - Dacite dome, north side of Mt. Adagdak.
46AC175   - Dacite summit dome, Mt. Adagdak (Coats, 1952).
76AMm204   - Dacite dome, north side of Mt. Adagdak.
76AMm207   - Dacite dome, top of Mt. Adagdak.

Andrew Bay

76AMm217A   - Andesite boulder in tuff-breccia, east side of Andrew Bay.
76AMm218   - Andesite dome, east side of Andrew Bay.
AD72   - Massive andesite flow in sea-cliff 1 km north of Andrew
            Lake (Marsh, 1976).

Moffett

46AC158   - Basalt from parasitic cone approximately 0.7 km east of
            summit (Coats, 1952).
46AC48   - Basalt flow in sea cliff 1.7 km northeast of summit of
            parasitic cone (Coats, 1952).
AD36   - Basalt flow from parasitic cone, bottom flow in sea cliff
            2 km NW of NW corner of Andrew Lake (Marsh, 1976).
AD39   - Basalt flow from parasitic cone of Mt. Moffett (Marsh, 1976).
46AC174   - Andesite dome 3.2 km south of summit of Mt. Moffett (Coats, 1952
AD44   - Andesite flow exposed along beach 3 km NW of NW corner of
            Andrew Lake (Marsh, 1976).
46AC275   - Andesite flow from composite cone 1.5 km NNW of summit of
            Mt. Moffett (Coats, 1952).
AD49   - Andesite flow 4 km NW of NW corner of Andrew Lake (Marsh, 1976).
76AMm222   - Andesite dome east side of Mt. Moffett.

* Samples 46AC36, 46AC21, 46AC13, 46AC175, 46AC158, 46AC48, 46AC174,
  46AC275 from Coats, 1952. Samples AD14, AD72, AD36, AD39, AD44,
  and AD49 from Marsh, 1976.

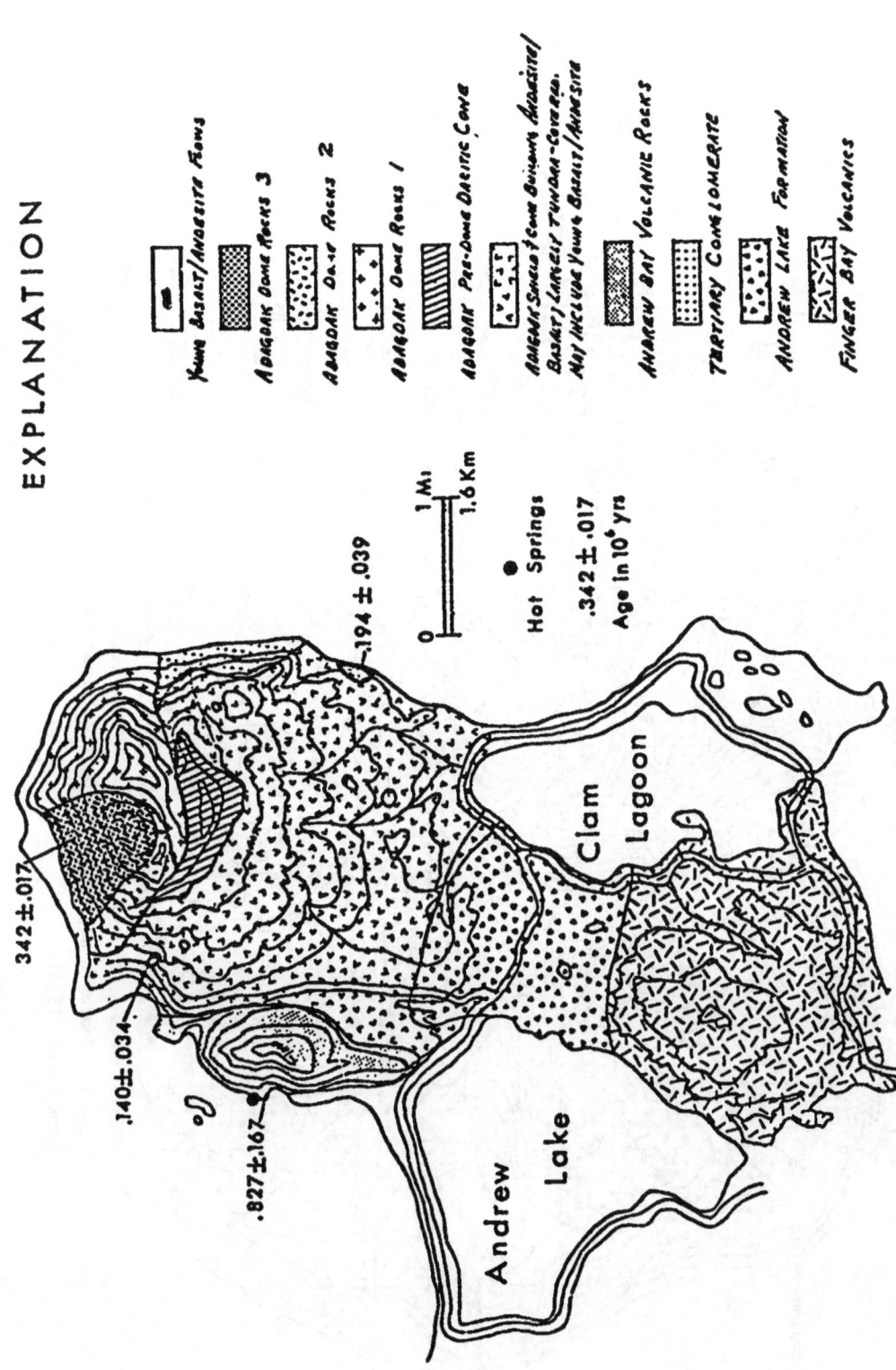

EXPLANATION

Young Basalt/Andesite Flows

Adagdak Dome Rocks 3

Adagdak Dome Rocks 2

Adagdak Dome Rocks 1

Adagdak Pre-Dome Dacitic Cone

Adagdak Strato (Cone Building Andesitic/
Basalt Lavas) Includes Tuyidak-Covered.
May Include Young Basalt/Andesita

Andrew Bay Volcanic Rocks

Tertiary Conglomerate

Andrew Lake Formation

Finger Bay Volcanics

.194 ± .039

342 ± .017

.140 ± .034

.827 ± .167

Clam Lagoon

Andrew Lake

0   1 Mi.
     1.6 Km

● Hot Springs

.342 ± .017
Age in 10⁶ yrs

Figure 1. Generalized geologic map of Adagdak and Andrew Bay volcanoes.

67

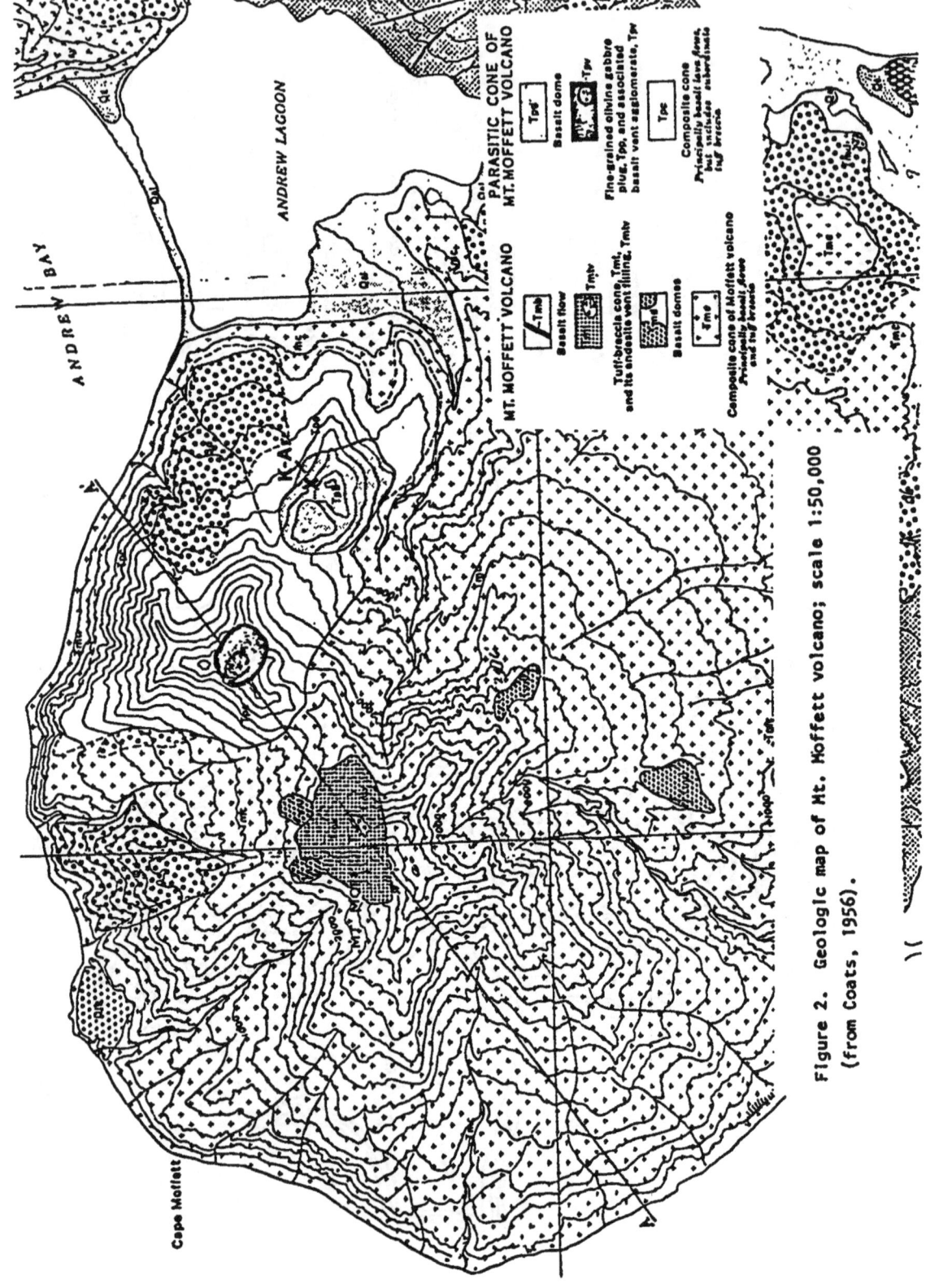

Figure 2. Geologic map of Mt. Moffett volcano; scale 1:50,000 (from Coats, 1956).

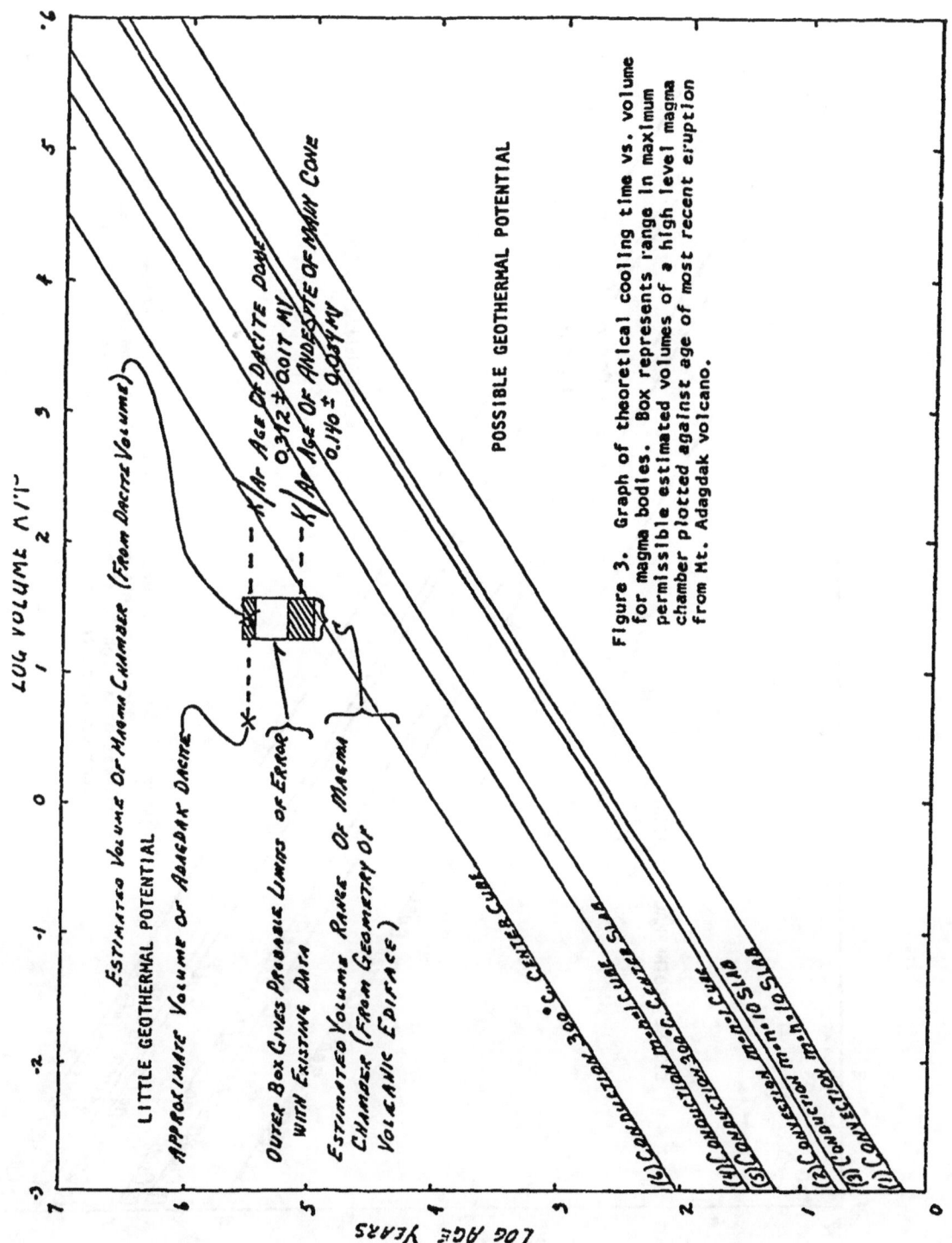

Figure 3. Graph of theoretical cooling time vs. volume for magma bodies. Box represents range in maximum permissible estimated volumes of a high level magma chamber plotted against age of most recent eruption from Mt. Adagdak volcano.

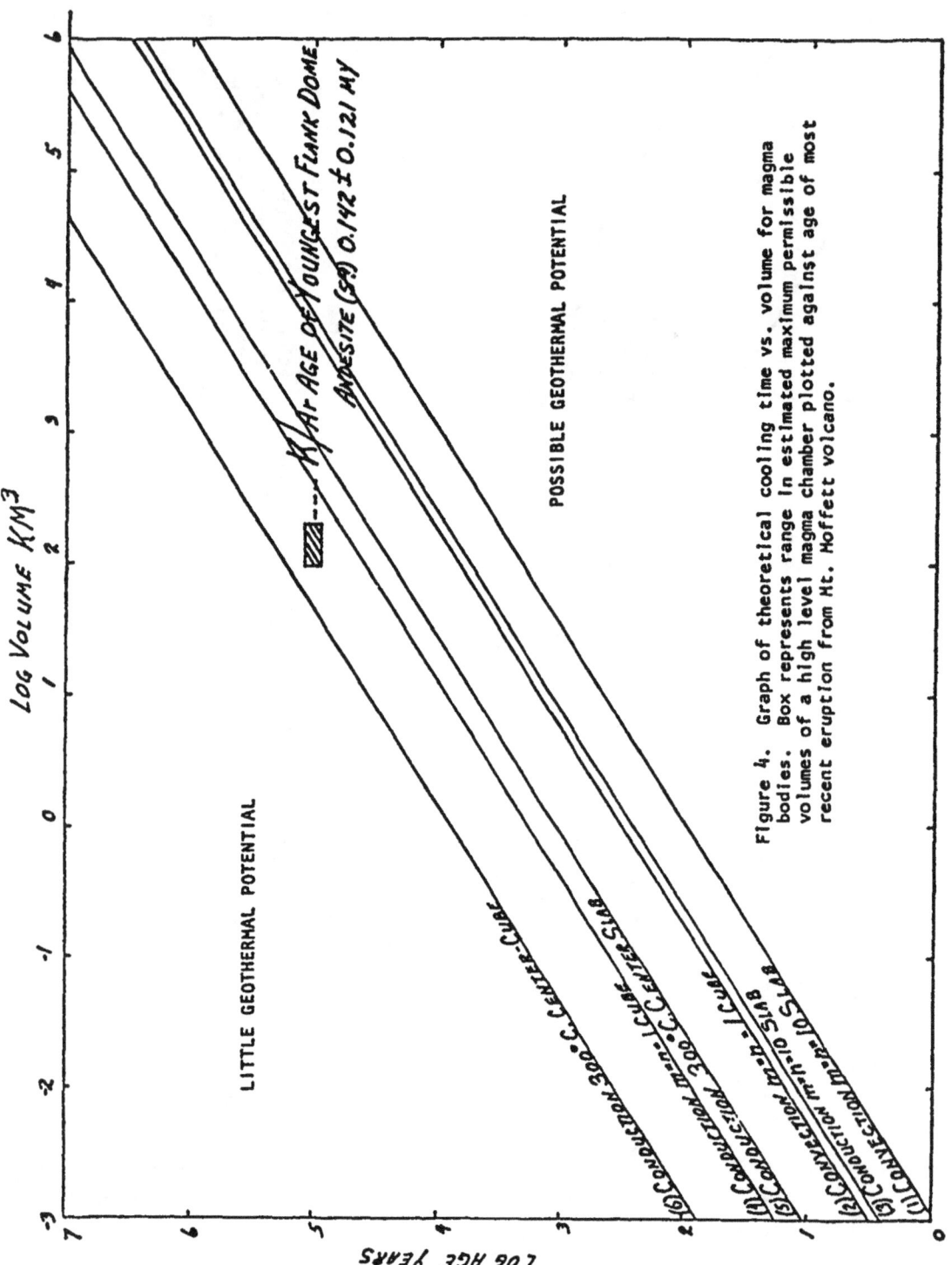

Figure 4. Graph of theoretical cooling time vs. volume for magma bodies. Box represents range in estimated maximum permissible volumes of a high level magma chamber plotted against age of most recent eruption from Mt. Moffett volcano.

Appendix B

THIN-SECTION DESCRIPTIONS, SAMPLES FROM ANDREW BAY
HOT SPRINGS AND KIKUGU WARM SPRINGS AREAS,
ADAK ISLAND

Date. 25 October 1982.
Collecor: Jim Brophy.

Field Notes:
    Original number: 6.
    Locality: Andrew Bay Hot Springs.
    Occurrence: Grab-sample from the Fe^{++}-stained highly brecciated zone at the northernmost rock exposure (Brophy).

Questions: (1) Classification, (2) description of alteration.

Hand specimen description: Limonite-stained fine-grained igneous rock.

Microscopic study for classification:
    Petographer: J. A. Whelan.
    Texture: Porphyritic.
    Secondary structure: Brecciation, hydrothermal alteration.
    Mineralogy:
      Phenocrysts:
        -Chlorite pseudomorphic after euhedral hornblende 25%, corroded. Up to +3 mm long.
        -Plagioclase, euhedral, zoned, 15%; some zones partially replaced by chlorite, sericite, and calcite. Up to +2 mm in size. Fractured.
      Groundmass:
        -Plagioclase, euhedral to anhedral. 40%. Up to 0.05 mm in size.
        -Quartz, anhedral, 10%, 0.05 mm.
        -Calcite (?), 4%.
        -Pyrite, 1%, 0.01 mm.
    Special features: Goethite coats the walls of a 1-mm fracture in this rock.

Classification: Propylitically altered andesite porphyry breccia.

Date: 20 October 1982.
Collector: Jim Brophy.

Field notes:
    Original number: 7.
    Locality: Andrew Bay Hot Springs.
    Occurrence: Taken from south of the two small peninsulars in the hot spring
       area. Seemingly the least altered rock in the area.

Question: Classification.

Hand specimen description: Andesite porphyry; grey; weathers brown; groundmass
       grey. About 10% plagioclase phenocrysts, 1 to 2 mm;
       about 10% hornblende phenocrysts, 1 to 2 mm.

Microscopic study for classification:
    Petrographer: J. A. Whelan.
    Texture: Porphyritic.
    Original structure: Flow.
    Secondary structure: Slight deuteric or hydrothermal alteration.
    Mineralogy:
      Phenocrysts:
        -Plagioclase (10%), euhedral fractured frequently showing several crystallogra-
        phic orientations. Zoned overgrowths. Cores An83-86 based on combined
        albite-carlsbad twinning.
        -Hornblende (5%); euhedral, frequently with rims of fine-grained anhedral
        magnetite.
      Groundmass:
        Average grain size under 0.05 mm.
        -Feldspar seemingly untwinned, cloudy, 60%.
        -Quartz: anhedral, clear, 20%.
        -Calcite (?), 5%.

Classification: Hornblende andesite porphyry.

Origin of rock: Flow.

Date: 26 October 1982.
Collector: Jim Brophy.

Field notes;
    Original number: 8.
    Locality: Andrew Bay Hot Springs.
    Occurrence: Altered rock from hot spring area.

Questions: (1) Classification, (2) description of alteration.

Hand specimen description: Fine-grained dense grey-green rock. Weathers tan; on sawed surface, grey. A few fragments or plagioclase phenocrysts are visible, up to about 6 mm. A very few green euhedral hornblende phenocrysts (1 mm) cut by a few irregular epidote veinlets up to 1 mm. One fracture surface covered with a black mineralization (oxides?).

Microscopic study for classification:
    Petographer: J. A. Whelan.
    Texture: Porphyritic.
    Original structure: Flow.
    Secondary structure: Hydrothermal alteration.
    Mineralogy:
        Phenocrysts:
            -Plagioclase feldspar (25%), euhedral, zoned. A poor determination on albite twins gave An50. Generally about 0.5 mm, but some up to several mm. Some sericitization and calcite in phenocrysts. Fractured.
            -Epidote and chlorite after euhedral hornblende (25%). About 1 mm long. Fine euhedral to anhedral pyrite (19%) associated with the epidote but also disseminated throughout the rock.
        Groundmass:
            Appears to be a mat of small euhedral plagioclase (40%) with interstitial epidote (10%).

Origin of rock: (1) Volcanism, (2) brecciation and hydrothermal alteration.

Date: 26 October 1982.
Collector: Jim Brophy.

Field notes:
    Original number: 9.
    Location: Andrew Bay Hot Springs.
    Occurrence: Most altered rock from sea-cut beach area of the hot spring.

Questions: (1) Classification, (2) description of alteration.

Hand specimen description: Grey porphyry with small (1 mm) hornblende
                         phenocrysts. Weathers tan. Black Mn oxide staining.

Microscopic study for classification:
    Petrographer: J. A. Whelan.
    Texture: Porphyritic.
    Original structure: Flow.
    Secondary structure: Hydrothermal alteration.
    Mineralogy:
        Phenocrysts:
            -Plagioclase (15%). Euhedral about 4 mm. Corroded, partially replaced by epidote.
            -Epidote (20%) pseudomorphic after euhedral hornblende, corroded, about 0.6 mm.
        Groundmass:
            -Plagioclase, about 0.03 mm, 60%: with fine epidote, 5%.
            -Pyrite in anhedral aggregates up to 0.6 mm closely associated with the epidote pseudomorphs.

Classification: Hydrothermally altered andesite porphyry.

Origiin of rock: (1) Volcanic activity, (2) hydrothermal alteration.

Date: 19 October 1982.
Collector: Jim Brophy.

Location: Kiguga Warm Springs.
Original number: 3.

Questions: (1) Classification. (2) type and degree of alteration.

Hand specimen description: Hydrothermally altered breccia zone. Note FeS$_2$ mineralization (Brophy). Well-cemented, very fine-grained anhedral to euhedral pyrite. This thin section is of one rock type, presumably a rock fragment in the breccia. A sample of the cement (numbered 3A) is being thinsectioned. Sample 3A is fine-grained, dense (barite is suspected), and contains limonite pseudomorphs after pyrite. (Slide 3A cement is chalcedony).

Microscopic study for classification:
    Petographer: J. A. Whelan.
    Texture: Microporphyritic.
    Mineralogy:
        Primary:
            Groundmass:
                -Glass: 65%, 0.01 mm; containing 5% (of the total rock), shards 0.01 mm by 0.002 mm. Appears to have been K-feldspathized, followed by sericitization.
                -K-feldspar: 5%, euhedral. 0.2 mm by 0.05 mm, sericitized.
                -Quartz: 5%, anhedral, 0.15 mm by 0.05 mm.
            Microphenocrysts:
                -K-feldspar: 20%, euhedral from 0.25 to 1.0 mm. Fresh to almost totally sericitized.
                -Plagioclase; 4%, up to 0.6 mm by 0.2 mm. Badly sericitized, shows oscillatory zoning.
                -Quartz: 1%, anhedral, unaltered, up to 0.3 mm.
        Secondary minerals: K-feldspar (?), sericite.
        Introduced minerals: Pyrite, anhedral to euhedral, approximately 0.2 mm.
    Special Features: This appears to be an ash-flow tuff which has undergone feldspathization and sericitization along with the introduction of pyrite.

Classification: Hydrothermally altered ash-flow tuff.

Origin: (1) Volcanic: (2) brecciation: (3) hydrothermal alteration.

Date: 19 October 1982.
Collector: Jim Brophy.

Location: Kiguga Warm Springs.
Original number: 4.
Occurrence: At the base of the cliff, close to the hydrothermally altered breccia
                zone (sample 3) (Brophy).

Questions: (1) Classification, (2) evidence of geothermal activity.

Hand specimen description: Limonite-stained fine-grained rock. On a sawed surface
                           it is a breccia, with light-tan corroded fragments, usually
                           about 3 mm but up to 25 mm in a grey cement.

Microscopic study for classification:
    Petrographer: J. A. Whelan.
    Texture: Breccia.
    Original structure: Quartz vein.
    Secondary structure: Brecciation and resealing.
    Mineralogy: Fragments are quartz (29%), anhedral, up to 0.25 mm and sericite
        (1%), on grain boundaries and in irregular patches up to 0.25
        mm in a matrix of very fine-grained (0.005 to 0.01 mm) quartz
        (68%) and sericite (2%). There are a few quartz veinlets up to 0.2
        mm cutting the rock. A trace of anhedral pyrite up to 0.2 mm is
        present.

Classification of rock: Brecciated quartz vein.

Origin of the rock: (1) Hydrothermal deposition of the quartz vein; (2) brecciation;
                    (3) hydrothermal healing of the breccia; (4) minor fracturing;
                    (5) healing of fractures with quartz.

Date: 19 October 1982.
Collector: Jim Brophy.

Field notes:
    Original number: 5.
    Location: Kigugua Warm Springs.
    Occurrence: Grab-sample from fractured and silicified wave-cut beach.

Questions: (1) Classification, (2) indications of geothermal activity.

Hand specimen description: Breccia consisting of chalky white corroded fragments (30%) up to 10 mm in grey matrix. Weathers tan.

Microscopic study for classification:
    Petrographer: J.A. Whelan.
    Texture: Brecciated.
    Mineralogy:
        Fragments: Consist of anhedral, unaltered fine quartz (70%) up to 0.2 mm in size and 30% sericitized euhedral K-feldspar up to 1 mm long. Cement is very fine-grained (0.01 mm) quartz. Trace of limonite after anhedral pyrite.

Classification: Quartz healed, brecciated rhyolitic tuff or tuffaceous sandstone.

Origin of rock: (1) Volcanic activity, (2) deposition of the tuff or tuffaceous sandstone, (3) brecciation, (4) geothermal solutions healing the breccia by the depostion of quartz. The same solutions sericitizing the K-feldspars and depositing pyrite.

Appendix C

TEMPERATURE LOG, ADAK ISLAND DRILL HOLES

On 12 January 1978, a maximum recording thermometer was lowered to total depth of 995 ft in Observation Hole 1. The thermometer, left at depth for 2.25 hours, gave a reading of 75°F.

Table C-1 is a record of temperatures from Observation Hole 2.

TABLE C-1. Temperature Logging Data, Observation Hole 2.

Depth, ft	T°F, down	T°F, up	Depth, m	T°C, down	T°C, up
0	41.2	40.5	0	5.1	4.7
50	41.0	39.7	15.2	5.0	4.3
100	41.9	41.1	30.5	5.5	5.1
150	43.5	43.0	45.7	6.4	6.1
200	45.4	44.6	61.0	7.4	7.0
250	47.3	46.4	76.2	8.5	8.0
300	49.5	50.0	91.5	9.7	9.8
350	51.4	50.2	106.7	10.8	10.1
400	53.1	52.2	122.0	11.7	11.2
450	56.5	54.5	137.2	13.6	12.5
500	57.9	57.0	152.4	14.4	13.9
550	59.7	58.5	167.7	15.4	14.7
600	60.6	60.8	182.9	15.9	16.0
650	62.6	62.2	198.2	17.0	16.8
700	65.8	64.5	213.4	18.8	18.0
750	68.5	66.4	228.7	20.3	19.1
800	70.3	68.0	243.9	21.3	20.0
850	72.7	72.5	259.1	22.6	22.5
900	74.7	75.0	274.3	23.7	23.9
950	77.0	76.6	289.6	25.0	24.8
1000	79.0	79.7	304.9	26.1	26.5
1050	81.3	81.5	320.1	27.4	27.5
1100	83.1	83.3	335.4	28.4	28.5
1150	85.5	85.8	350.6	29.7	29.9
1200	87.4	88.0	365.9	30.8	31.1
1250	90.0	90.5	381.1	32.2	32.5
1300	91.9	92.8	396.3	33.3	33.8
1350	94.4	94.8	411.6	34.6	34.9
1400	96.2	97.0	426.8	35.6	36.1
1450	98.4	99.3	442.1	36.9	37.4
1500	100.4	100.9	457.3	38.0	38.3
1550	102.9	103.3	472.6	39.4	39.6
1600	105.1	105.6	487.8	40.6	40.9
1650	107.2	107.4	503.0	41.8	41.9
1700	109.6	110.1	518.3	43.1	43.4
1750	111.8	111.9	533.5	44.3	44.4
1800	114.1	112.8	548.8	45.6	44.9
1850	116.4	115.3	564.0	46.9	46.3
1900	118.9	117.7	579.3	48.3	47.6
1925	118.6	...	586.9	48.1	...

Notes:
1. Logged by R. Clodt and J. Neffew 24 October 1978.
2. Average Gradients: 50-1900 ft-4.2°F/100 ft; 15.2-586.9 m-7.6°C/100 m.

Appendix D

GRAVITY AND MAGNETIC FACTS, NORTHERN ADAK

Station ID	Latitude[a]		Longitude[a]		Elevation, ft	Gravity,[b] mgal	Terr. corr.,[c] 2.0 gm/cc			Complete Bouguer anomalies,[d] gm/cc			Corrected magnetics, gammas
							Inner zone	Outer zone	Total	2.00	2.67	2.40	
LORAN	51	59.60	176	36.63	154.37	981417.41	1.33	.12	1.45	183.3	182.5	182.8	48098
1	51	59.53	176	36.67	152.55	981417.86	1.45	.12	1.57	183.9	183.1	183.4	48441
2	51	59.43	176	36.57	181.88	981415.66	1.44	.12	1.56	183.8	182.8	183.2	48290
3	51	59.35	175	36.43	229.08	981411.96	1.91	.12	2.03	183.9	182.7	183.2	47787
4	51	59.25	175	36.52	194.36	981414.57	2.34	.11	2.45	184.7	183.9	184.2	47284
5	51	59.17	176	36.68	199.19	981415.05	2.73	.12	2.85	186.1	185.3	185.6	47397
6	51	59.10	176	36.83	221.65	981414.55	2.31	.13	2.44	186.8	185.7	186.1	47605
7	51	58.95	176	36.88	234.34	981414.36	1.90	.15	2.05	187.3	186.0	186.5	47984
8	51	58.85	176	36.80	220.80	981414.52	1.80	.16	1.95	186.6	1853	185 ?	47833
9	51	58.73	176	36.73	283.19	981410.12	1.68	.18	1.85	186.5	184.7	185 ?	47761
10	51	58.60	176	36.68	282.11	981409.81	1.47	.17	1.65	186.1	184.3	18? ?	47973
11	51	58.47	176	36.70	281.49	981409.18	1.70	.18	1.88	185.9	184.1	184.8	47588
12	51	58.30	176	36.58	313.81	981406.24	.93	.20	1.13	184.6	182.3	183.3	47656
13	51	58.12	176	36.50	314.22	981406.14	.55	.21	.76	184.5	182.1	183.0	48329
14	51	57.97	176	36.50	283.70	981408.32	.43	.20	.63	184.7	182.4	183.3	48388
15	51	57.85	176	36.52	262.41	981409.89	.46	.22	.68	185.0	183.0	183.8	48458
16	51	57.70	176	36.52	244.38	981411.11	.28	.22	.50	185.0	183.1	183.9	48533
17	51	57.53	176	36.52	224.49	981412.59	.26	.21	.47	185.3	183.6	184.3	48402
18	51	57.38	176	36.43	229.56	981412.34	.23	.20	.43	185.6	183.8	184.5	48328
19	51	57.25	176	36.38	225.31	981412.98	.45	.17	.62	186.4	184.6	185.3	48367
20=M8	51	57.20	176	36.32	219.64	981413.57	.30	.17	.47	186.5	184.8	185.4	48552
M9	51	57.18	176	36.02	211.19	981414.90	.22	.18	.40	187.2	185.5	186.2	...
21	51	57.18	176	36.00	208.88	981414.85	.26	.18	.44	187.0	185.4	186.0	48974
22	51	57.15	176	35.67	204.74	981415.07	.21	.17	.38	186.9	185.3	186.0	49584
23	51	57.12	176	3	188.85	981416.42	.21	.15	.36	187.2	185.7	186.3	48360
M8F	51	57.00	176	3:	142.07	981420.80	.17	.14	.31	188.5	187.4	187.9	...
M8G	51	57.02	176	35	139.03	981420.92	.16	.15	.30	188.4	187.3	187.8	...
24	51	56.92	176	35.32	93.44	981425.59	.19	.15	.33	190.1	189.4	189.7	48533
25=EA	51	56.67	176	35.30	41.68	981431.40	.08	.17	.25	192.7	192.4	192.5	...
BS2	51	56.73	176	35.07	10.29	981432.84	.19	.19	.38	192.0	192.0	192.0	...
BS3	51	56.65	176	35.08	8.28	981433.74	.12	.18	.30	192.8	192.8	192.8	...
26	51	56.57	176	35.30	13.07	981433.80	.15	.17	.33	193.4	193.4	193.4	48311
27	51	56.43	176	35.38	11.98	981434.76	.14	.17	.31	194.4	194.4	194.4	47747
28	51	56.27	176	35.38	13.32	981435.40	.22	.15	.37	195.4	195.5	195.5	48106
29=BS6	51	56.30	176	35.52	45.00	981434.28	.15	.15	.29	196.4	196.1	196.2	...
30	51	56.27	176	35.68	63.39	981432.38	.23	.16	.39	195.9	195.5	195.6	47826
M2	51	56.18	176	35.82	132.73	981428.43	.60	.14	.74	197.1	196.3	196.6	...

See footnotes at end of table.

Station ID	Lattitude[a]		Longitude[a]		Elevation, ft	Gravity,[b] mgal	Terr. corr.,[c] 2.0 gm/cc			Complete Bouguer anomalies,[d] gm/cc			Corrected magnetics, gammas
							Inner zone	Outer zone	Total	2.00	2.67	2.40	
31	51	56.25	176	35.92	72.73	981431.62	.23	.16	.39	195.8	195.3	195.5	47699
M3	51	56.33	176	36.23	99.50	981428.46	.40	.17	.57	194.5	193.9	194.1	...
32	51	56.38	176	36.45	68.42	981429.77	.12	.22	.33	193.4	192.9	193.1	48093
33	51	56.55	176	36.70	75.26	981426.42	.21	.23	.44	190.4	189.9	190.1	48655
M4	51	56.55	176	36.65	102.80	981424.39	.78	.21	1.00	190.8	190.2	190.5	...
GEN	51	56.67	176	36.55	80.94	981425.41	.20	.23	.42	189.6	189.0	189.2	...
34	51	56.73	176	36.65	83.65	981424.26	.24	.23	.47	188.5	188.0	188.2	49575
35	51	57.02	176	36.52	36.43	981426.24	.28	.24	.51	186.9	186.8	186.8	47850
36	51	57.12	176	36.75	20.13	981426.69	.38	.26	.64	186.2	186.3	186.3	48155
37	51	57.13	176	36.90	22.52	981426.27	.59	.26	.85	186.2	186.3	186.2	50191
38	51	57.23	176	37.05	20.99	981425.30	.95	.26	1.21	185.3	185.5	185.4	47772
39	51	57.35	176	37.22	20.18	981424.53	1.44	.27	1.71	184.8	185.2	185.0	48040
40	51	57.47	176	37.40	19.73	981424.34	1.13	.32	1.45	184.1	184.5	184.3	47006
41	51	57.53	176	37.65	20.13	981424.44	.31	.33	.64	183.4	183.4	183.4	48184
42	51	57.43	176	38.02	22.43	981423.52	.12	.32	.44	182.5	182.5	182.5	48380
43	51	57.38	176	38.30	21.84	981423.29	.11	.41	.51	182.4	182.4	182.4	47043
44	51	57.32	176	38.50	20.82	981423.09	.10	.45	.55	182.3	182.3	182.3	47462
45	51	57.27	176	38.75	30.67	981423.23	.13	.47	.60	183.2	183.2	183.2	47415
46	51	57.20	176	39.02	28.90	981422.23	.14	.51	.65	182.3	182.2	182.2	49157
47	51	57.15	176	39.27	32.20	981422.06	.20	.60	.80	182.5	182.5	182.5	47298
48	51	57.12	176	39.50	32.12	981422.00	.42	.65	1.07	182.8	182.9	182.8	47637
49	51	57.10	176	39.78	30.32	981421.40	1.48	.67	2.16	183.2	183.6	183.5	47962
50	51	56.97	176	39.87	20.80	981421.96	1.97	.70	2.67	183.8	184.5	184.2	47361
51	51	56.80	176	39.77	23.27	981422.91	1.28	73	2.01	184.5	185.0	184.8	47773
52	51	56.65	176	39.78	18.62	981423.82	1.46	.78	2.24	185.5	186.1	185.9	47862
53	51	56.50	176	39.78	20.30	981424.77	1.23	.80	2.03	186.6	187.1	186.9	47922
54	51	56.45	176	39.42	19.01	981425.45	.40	.72	1.12	186.4	186.6	186.5	52085
55	51	56.40	176	39.08	22.15	981426.16	.24	.64	.87	187.1	187.2	187.2	47831
56	51	56.33	176	38.88	25.90	981426.36	.17	.56	.73	187.5	187.6	187.5	48044
57	51	56.20	176	38.70	25.61	981427.74	.12	.52	.64	189.0	189.0	189.0	52311
58	51	56.12	176	38.58	22.35	981428.80	.17	.51	.68	190.0	190.0	190.0	47870
59	51	55.92	176	38.38	19.66	981431.14	.14	.47	.61	192.4	192.4	192.4	46841
60	51	55.77	176	38.38	18.82	981432.31	.15	.47	.62	193.7	193.8	193.7	48042
61	51	55.62	176	38.47	18 84	981433.77	.19	.48	67	195.4	195.5	195.5	47774
62	51	55.35	176	38.58	47.70	981433.83	.33	.47	.79	198.0	197.9	197.9	47427
63	51	55.15	176	38.53	139.12	981430.00	.30	.36	.66	200.6	199.6	200.0	47894
64	51	54.95	176	38.50	180.67	981427.91	.26	.33	59	201.6	200.2	200.8	48602
65	51	54.75	176	38.32	165 83	981429.91	.19	.31	50	202.8	201.5	202.0	47687
66	51	54.62	176	38.15	151.49	981431.66	.19	.25	.44	203.6	202.5	203.0	48488
67=AU2	51	54.52	176	37.95	148.83	981432.53	.12	.24	.36	204.4	203.3	203.7	47293
68	51	54.53	176	37.72	191.18	981430.22	.22	.22	.44	205.1	203.6	304.2	47805
E19	51	54.63	176	37.50	162.24	981432.03	.16	.22	.37	204.7	203.4	203.9	...
69	51	54.62	176	37.47	166.91	981432.19	.16	.22	.38	205.2	203.9	204.4	46654
70	51	54.72	176	37.20	202.09	981430.29	.15	.17	.32	205.5	203.9	204.5	47864
71	51	54.62	176	37.05	244.22	981427.87	.26	.16	.41	206.2	204.3	205.0	48310
BS19	51	54.58	176	36.78	169.75	981430.05	.74	.16	.90	203.8	202.7	203.1	...
BS18	51	54.70	176	36.45	165.02	981432.97	.29	.17	.45	205.8	204.5	205.0	...
E14	51	55.07	176	36.25	279.88	981425.00	.38	14	.52	205.2	203.0	203.9	...
72	51	54.68	176	36.20	167.34	981433.20	.13	.13	.26	206.0	204.7	205.2	47616
73	51	54.70	176	35.87	177.10	981432.47	.29	.13	.42	206.1	204.7	205.3	48109
74	51	54.70	176	35.57	163.06	981433.47	.33	.11	.43	206 1	204.9	205.4	47898
75	51	54.72	176	35.32	82.41	981438.50	.22	.11	.33	205.5	204.9	205.2	48984

See footnotes at end of table.

Station ID	Lattitude[a]	Longitude[c]	Elevation, ft	Gravity,[b] mgal	Terr. corr.,[c] 2.0 gm/cc			Complete Bouguer anomalies,[d] gm/cc			Corrected magnetics, gammas
					Inner zone	Outer zone	Total	2.00	2.67	2.40	
E8	51 54.82	176 35.32	83.13	981438.84	.20	.11	.31	205.7	205.1	205.4	...
76	51 54.85	176 35.12	67.99	981438.81	.19	.12	.31	204.6	204.1	204.3	48497
77	51 55.00	176 35.00	33.50	981441.23	.13	.13	.25	204.4	204.2	204.3	48395
E10	51 55.12	176 35.12	10.15	981443.36	.21	.13	.34	204.9	204.9	204.9	...
78	51 55.13	176 35.15	12.52	981442.91	.23	.13	.36	204.6	204.6	204.6	48279
79	51 55.50	176 35.10	11.33	981441.06	.36	.15	.51	202.2	202.3	202.3	48035
80	51 55.35	176 35.15	10.29	981441.80	.64	.14	.78	203.4	203.6	203.5	48196
81	51 55.63	176 34.98	9.12	981440.14	.17	.15	.32	200.8	200.8	200.8	48810
82	51 55.88	176 34.97	13.29	981438.71	.26	.14	.40	199.4	199.4	199.4	48651
83	51 56.08	176 35.20	9.59	981437.63	.25	.15	.40	197.7	197.8	197.8	49085
84	51 56.18	176 35.18	16.41	981435.88	.18	.15	.33	196.2	196.2	196.2	48313
85	51 54.43	176 38.05	111.29	981435.00	.14	.26	.40	204.5	203.7	204.0	48625
86	51 54.22	176 37.98	75.33	981437.67	.15	.27	.42	205.0	204.5	204.7	47907
87	51 54.17	176 37.80	35.94	981440.28	.17	.28	.45	205.0	204.9	204.9	48754
88	51 54.18	176 37.57	44.25	918439.75	.15	.27	.41	205.0	204.8	204.9	48948
C9	51 54.25	176 37.43	87.31	981437.15	.15	.24	.39	205.2	204.6	204.9	...
89	51 54.23	17 0 37.40	89.28	981437.16	.14	.24	.38	205.4	204.8	205.0	48799
90	51 54.27	17 37.17	126.25	981434.66	.23	.17	.40	205.4	204.5	204.8	49881
91	51 54.38	176 37.05	168.18	981432.46	.24	.16	.40	205.9	204.6	205.1	48224
E1	51 54.42	176 37.02	164.77	981432.56	.21	.16	.38	205.7	204.4	204.9	...
E3	51 54.52	176 36.75	181.56	981431.74	.29	.16	.45	206.0	204.6	205.1	...
E4	51 54.58	176 36.58	137.82	981434.67	.32	.16	.48	205.8	204.8	205.2	...
E5	51 54.60	176 36.28	187.28	981431.76	.28	.12	.40	206.2	204.7	205.3	...
E6	51 54.62	176 36.10	189.47	981431.28	.43	.12	.55	206.0	204.6	205.1	...
92	51 54.55	176 36.97	223.35	981429.22	.14	.16	.29	206.1	204.3	205.0	48168
93	51 56.75	176 35.38	31.00	981430.32	.14	.18	.31	190.8	190.6	190.7	48015
94	51 56.73	176 35.65	44.09	981429.58	.08	.20	.28	191.0	190.7	190.8	48070
OFF	51 56.85	176 35.95	67.70	981426.01	.12	.18	.30	188.9	188.4	188.6	...
95	51 56.83	176 36.28	95.74	981424.10	.12	.16	.29	188.9	188.1	188.4	52538
RUN	51 56.82	176 36.40	70.53	981423.62	.49	.20	.69	187.1	186.7	186.9	...
96	51 56.88	176 34.98	12.21	918431.52	.31	.17	.48	190.7	190.7	190.7	48191
97	51 57.02	176 34.85	5.68	981430.91	.36	.18	.54	189.5	189.6	189.6	47928
98	51 57.13	176 34.77	19.05	981428.71	.42	.18	.61	188.1	188.2	188.1	47902
99	51 57.27	176 34.32	6.92	981427.99	.38	.18	.56	186.3	186.4	186.4	48324
100	51 57.22	176 34.10	7.18	981428.42	.27	.18	.45	186.7	186.8	186.8	48486
101	51 57.08	176 33.90	7.85	981429.61	.12	.17	.29	188.0	188.0	188.0	48436
102	51 56.95	176 33.88	8.39	981430.98	.08	.16	.24	189.5	189.5	189.5	48286
103	51 56.78	176 33 83	3.41	981432.69	.06	.17	.23	191.2	191.2	191.2	48323
104	51 56.65	176 33.77	3.32	981433.98	.07	.14	.21	192.6	192.7	192.7	48437
105	51 56.50	176 33.67	2.24	981435.42	.04	.13	.17	194.2	194.2	194.2	49057
106	51 56.37	176 33.55	18.04	981435.96	.06	.12	.18	196.0	195.9	195.9	48691
107	51 56.22	176 33.40	16.84	981437.69	.05	.11	.16	197.8	197.7	197.8	49691
108	51 56.12	176 33.27	18.15	981438.68	.06	.11	.16	199.1	199.0	199.0	48447
109	51 56.05	176 33.15	22.72	981439.07	.06	.10	.16	199.9	199.7	199.8	48611
110	51 55.88	176 33.07	22.52	981439.08	.08	.10	.18	200.1	200.0	200.0	48693
111	51 55.73	176 33.02	8.49	981441.28	.13	.10	.24	201.6	201.6	201.6	48518
112	51 55.75	176 33.27	5.70	981441.84	.03	.11	.14	201.9	201.9	201.9	48291
113	51 55.67	176 33 52	10.35	981442.46	.16	.10	.27	203.1	203.1	203.1	48459
114	51 55.55	176 33.72	9.87	981442.57	30	.10	.40	203.5	203.5	203.5	47856
115	51 55.33	176 33 82	8.23	981443.21	.07	.10	.17	204.1	204.1	204.1	47861
116	51 55.18	176 33.95	9.85	981442.88	.04	.11	.15	204.1	204.0	204.0	48246
117	51 55.07	176 34.35	47.17	981440.19	.08	.10	.18	204.1	203.8	203.9	47927

See footnotes at end of table

Station ID	Lattitude[a]		Longitude[a]		Elevation, ft	Gravity,[b] mgal	Terr. corr.,[c] 2.0 gm/cc			Complete Bouguer anomalies,[d] gm/cc			Corrected magnetics, gammas
							Inner zone	Outer zone	Total	2.00	2.67	2.40	
118	51	55.13	176	34.55	20.62	981441.95	.09	.11	.20	204.0	203.9	203.9	48393
119	51	55.08	176	34.78	11.34	981442.05	.12	.13	.25	203.6	203.6	203.6	48244
120	51	54.12	176	37.13	40.25	981439.60	.18	.18	.37	204.6	204.4	204.5	48588
121	51	53.98	176	37.17	14.55	981440.00	.06	.20	.26	203.4	203.3	203.3	48749
122	51	53.83	176	37.25	13.64	981439.53	.04	.20	.24	203 0	203.0	203.0	48090
G3	51	53.80	176	37.33	11.06	981439.22	.06	.21	.27	202.6	202.6	202.6	...
123	51	53.68	176	37.32	13.78	981439.39	.03	.20	.23	203.1	203.1	203.1	48258
G4	51	53.60	176	37.32	14.17	981439.27	.03	.20	.23	203.1	203.1	203.1	...
124	51	53.48	176	37.45	12.49	981439.64	.05	.24	.29	203.6	203.6	203.6	45997
125	51	53.30	176	37.58	19.33	981439.51	.04	.24	.28	204.2	204.2	204.2	48221
126	51	53.15	176	37.60	17.65	981440.56	.06	.24	.30	205.4	205.4	205.4	48238
127	51	52.93	176	37.57	37.80	981439.94	.18	.21	.38	206.6	206.4	206.5	50012
J15	51	52.83	176	37.77	210.03	981427.33	2.47	.15	2.62	208.1	207.2	207.6	...
128	51	52.83	176	37.57	48.34	981439.50	.33	.20	.53	207.1	206.9	207.0	49974
J14	51	52.77	176	37.65	37.65	981439.57	.18	.21	.39	206.4	206.2	206.3	...
J13	51	52.62	176	37.67	31.95	981439.79	.02	.21	.23	206.3	206.1	206.2	...
129	51	52.62	176	37.65	35.74	981439.69	.03	.21	.24	206.5	206.3	206.3	48645
130	51	52.48	176	37.62	26.16	981441.03	.04	.21	.26	207.4	207.3	207.3	48141
J12	51	52.43	176	37.62	23.27	981441.50	.39	.22	.61	208.1	208.1	208.1	...
131	51	52.32	176	37.77	20.44	981442.49	.45	.22	.67	209.1	209.2	209.1	48133
132	51	52.15	176	37.82	31.29	981442.28	.17	.21	.38	209.6	209.5	209.5	48130
133	51	52.00	176	37.98	21.61	981443.60	.06	.22	.28	210.4	210.3	210.3	47846
134	51	52.00	176	37.75	21.10	981444.16	.11	.22	.33	211.0	210.9	210.9	47882
135	51	51.97	176	38.28	22.87	981442.60	.04	.24	.28	209.5	209.4	209.4	48279
136	51	51.97	176	38.60	23.62	981442.63	.04	.24	.29	209.6	209.5	209.5	47957
137	51	52.08	176	38.48	29.68	981441.82	.05	.24	.30	209.0	208.9	209.0	47985
H7	51	52.15	176	38.43	29.96	981441.58	.05	.24	.29	208.7	208.6	208.6	...
138	51	51.77	176	38.87	12.92	981444.29	.09	.26	.35	210.9	210.9	210.9	48235
139	51	52.23	176	38.32	28.45	981441.71	.03	.24	.27	208.6	208.4	208.5	48485
H6	51	52.27	176	38.30	24.28	981441.72	.03	.24	.27	208.3	208.2	208.2	...
H5	51	52.42	176	38.25	28.35	981441.13	.03	.24	.27	207.7	207.6	207.7	...
140	51	52.43	176	38.23	31.47	981440.75	.02	.24	.25	207.5	207.3	207.4	47212
H4	51	52.60	176	38.18	32.08	981440.32	.03	.22	.25	206.9	206.7	206.8	...
141	51	52.65	176	38.17	38.23	981439.86	.04	.22	.26	206.8	206.6	206.7	48021
H3	51	52.77	176	38.10	40.95	981439.50	.06	.22	.28	206.5	206.2	206.3	...
142	51	52.80	176	38.10	49.00	981439.02	.11	.21	.32	206.5	206.2	206.3	48410
143	51	52.90	176	38.02	62.60	981439.46	.15	.20	.35	207.8	207.4	207.5	50173
144	51	53.03	176	38.00	53.68	981439.74	.18	.21	.39	206.3	206.0	206.1	47662
145	51	53.12	176	37.85	62.36	981438.20	.22	.22	.44	206.3	205.9	206.1	49658
146	51	51.67	176	39 12	18.09	981444.86	.25	.28	.53	212.1	212.2	212.2	48277
147	51	51.75	176	39.28	55.72	981442.47	.20	.25	.45	212.1	211.8	211.9	48211
148	51	51.93	176	39.38	40.12	981442.48	.50	.26	.76	211.1	211.0	211.0	47983
A10	51	52.03	176	39.30	35.74	981443.07	.15	.28	.43	210.9	210.8	210.8	...
J-12-5	51	52.07	176	39.50	56.01	981441.85	.25	.27	.51	211.1	210.8	210.9	...
149	51	52.13	176	39.48	47.41	981441.96	.28	.27	.55	210.6	210.4	210.4	48237
A12	51	52.37	176	39.40	48.84	981442.70	.34	.28	.62	211.1	210.9	211.0	...
150	51	52.35	176	39.55	80.94	981439.81	.36	.26	.62	210.5	210.0	210.2	48005
A13	51	52.55	176	39.42	28.66	981443.31	.26	.31	.57	210.1	210.0	210.0	...
151	51	52.55	176	39.63	48.82	981440.95	.45	.30	.74	209.3	209.1	209.2	47962
A14	51	52.60	176	39.67	89.19	981439.32	.27	.28	.54	210.1	209.5	209.8	...
152	51	52.68	176	39.68	54.10	981440.57	.77	.31	1.07	209.4	209.3	209.3	48196
L13	51	52.80	176	39.72	154.51	981434.78	.34	.26	.60	209.8	208.7	209.1	...

See footnotes at end of table.

Station ID	Lattitude[a]		Longitude[a]		Elevation, ft	Gravity,[b] mgal	Terr. corr.,[c] 2.0 gm/cc			Complete Bouguer anomalies,[d] gm/cc			Corrected magnetics, gammas
							Inner zone	Outer zone	Total	2.00	2.67	2.40	
153	51	52.83	176	39.63	78.58	981439.03	.58	.30	.88	209.1	208.7	208.9	50629
L12	51	52.93	176	39.52	118.51	981436.32	.31	.28	.59	208.7	207.9	208.2	...
154	51	52.92	176	39.30	40.71	981441.15	.50	.32	.82	208.4	208.4	208.4	48112
155	51	53.02	176	39.13	79.42	981438.23	.25	.29	.54	207.7	207.2	207.4	48809
L10	51	53.12	176	39.12	148.38	981433.30	.51	.29	.79	207.6	206.6	207.0	...
156	51	53.13	176	38.93	73.59	981438.40	.17	.28	.45	207.2	206.8	207.0	50897
L9	51	53.20	176	38.93	102.09	981435.77	.66	.27	.93	207.0	206.4	206.6	...
157	51	53.28	176	38.70	60.00	981437.78	.25	.29	.54	205.6	205.2	205.4	49407
L8	51	53.30	176	38.73	60.69	981437.93	.28	.29	.57	205.8	205.4	205.6	...
158	51	53.50	176	38.60	92.98	981435.24	.29	.27	.56	205.0	204.4	204.6	47641
159	51	53.63	176	38.53	110.71	981433.76	.34	.28	.62	204.6	203.9	204.2	48997
160	51	53.80	176	38.48	125.40	981432.84	.32	.27	.59	204.4	203.5	203.9	49505
B16	51	53.90	176	38.58	174.86	981430.05	.28	.26	.54	204.8	203.5	204.0	...
161	51	53.93	176	38.37	111.29	981434.60	.25	.27	.52	204.9	204.2	204.5	49964
162	51	54.05	176	38.10	53.17	981438.80	.15	.26	.41	204.9	204.5	204.7	52341
D2	51	54.10	176	37.93	20.55	981439.71	.17	.27	.44	203.5	203.5	203.5	...
163	51	54.15	176	38.12	24.86	981440.61	.33	.31	.64	204.8	204.8	204.8	48532
164	51	54.22	176	38.38	51.64	981438.37	.45	.35	.80	204.5	204.3	204.4	48136
165	51	54.32	176	38.45	83.96	981435.92	.24	.34	.58	203.9	203.3	203.6	48146
166	51	54.42	176	38.53	110.94	981433.66	.25	.34	.59	203.3	202.6	202.9	48012
167	51	54.58	176	38.60	97.77	981433.81	.49	.35	.84	202.6	202.0	202.2	47781
168	51	54.72	176	38.65	98.18	981433.12	.57	.38	.95	201.8	201.3	201.5	47947
C2	51	54.87	176	38.68	120.34	981431.36	.32	.38	.69	201.1	200.3	200.6	...
169	51	54.83	176	38.53	98.95	981433.04	.47	.37	.84	201.5	201.0	201.2	47413
C3	51	54.85	176	38.47	171.17	981428.60	.35	.33	.68	201.8	200.6	201.1	...
170	51	52.72	176	39.95	143.60	981435.81	.35	.26	.61	210.2	209.2	209.6	48020
A16	51	52.70	176	40.00	157.52	981435.21	.25	29	.54	210.5	209.4	209.8	...
171	51	52.75	176	40.22	161.27	981434.90	28	.29	.58	210.5	209.3	209.7	48294
172	51	52.67	176	40.43	169.56	981434.45	.31	.29	.60	210.7	209.4	210.0	48449
173	51	52.57	176	41.02	322.73	981426.23	.17	.27	.45	213.0	210.4	211.4	48724
174	51	52.67	176	41.05	344.18	981423.96	.22	.28	.50	212.1	209.3	210.4	48814
175	51	52.88	176	40.93	377.39	981421.99	.48	.29	.78	212.3	209.4	210.6	48349
176	51	53.08	176	40.97	504.84	981412.79	.85	.36	1.21	212.0	208.1	209.7	48469
177	51	53.25	176	40.93	572.54	981407.59	.55	.40	.95	210.9	206.4	208.2	48934
178	51	53.40	176	40.85	618.67	981403.69	.46	.43	.88	209.9	204.9	207.0	48921
179	51	53.90	176	38.55	170.02	981430.48	.24	.26	.50	204.9	203.6	204.1	48832
180	51	53.87	176	38.80	230.18	981426.24	.28	.25	.53	204.8	203.0	203.7	...
181	51	53.77	176	39.07	318.22	981420.15	29	.25	.54	204.9	202.4	203.4	48376
182	51	53.63	176	39.28	349.50	981418.70	.21	.29	.50	205.7	202.9	204.0	...
183	51	53.72	176	39.55	410.07	981414.42	.45	.31	.76	205.8	202.5	203.8	48365
184	51	53.58	176	39 57	389.24	981416.85	.28	.28	.55	206.8	203.6	204.9	48642
185	51	53.63	176	39.83	442.53	981413.06	.47	.32	.78	206.8	203.3	204.7	48896
186	51	53.68	176	40.10	538.84	981405.16	.89	.38	1.27	205.9	201.7	203.4	48441
187	51	53.70	176	40.32	596.22	981401.20	.77	.42	1.19	205.8	201.1	203.0	48498
188	51	53.58	176	40.43	596.78	981402.96	.55	.39	.94	207.5	202.7	204.6	48379
189	51	53.48	176	40.48	600.88	981403.68	.67	.39	1.06	208.8	204.0	205.9	48612
190	51	53.47	176	40.62	639.30	981402.19	.60	.41	1.02	209.9	204.8	206.8	48405
191	51	52.77	176	39.80	154.05	981434.80	.32	.25	.58	209.8	208.7	209.2	47630
192	51	52.87	176	39.73	131.34	981435.97	.46	27	.73	209.4	208.6	208.9	47762
193	51	53.00	176	39.65	159.28	981433.86	.47	27	.73	209 1	207.9	208.4	47985
194	51	53.13	176	39.83	267.42	981426.77	.49	.27	.76	209.2	207.2	208 0	47676
195	51	53.23	176	39.90	308.85	981424.04	.44	.27	.71	209.1	206.7	207.7	47878

See footnotes at end of table.

Station ID	Lattitude[a]	Longitude[a]	Elevation, ft	Gravity,[b] mgal	Terr. corr.,[c] 2.0 gm/cc			Complete Bouguer anomalies,[d] gm/cc			Corrected magnetics, gammas
					Inner zone	Outer zone	Total	2.00	2.67	2.40	
196	51 53.33	176 40.05	393.64	981418.11	.37	.32	.68	208.8	205.7	207.0	48630
197	51 53.42	176 40.15	437.61	981415.22	.38	.33	.71	208.9	205.3	206.8	49157
198	51 53.45	176 40.00	429.13	981415.60	.41	.32	.73	208.6	205.2	206.6	47882
199	51 53.57	176 39.80	398.61	981416.41	.27	.28	.56	207.0	203.8	205.1	4894!
200	51 52.33	176 39.68	92.69	981439.52	.29	.25	.54	210.9	210.3	210.6	48053
201	51 52.28	176 39.77	126.66	981437.47	.16	.23	.39	211.1	210.2	210.6	47903
202	51 52.17	176 39.87	145.37	981436.23	17	.22	.39	211.4	210.2	210.7	48303
203	51 52.07	176 39.97	157.11	981435.39	.15	.25	.39	211.5	210.2	210.7	48124
204	51 51.95	176 40.20	146.91	981436.02	.23	.22	.45	211.6	210.5	211.0	48124
205	51 51.87	176 40.38	184.26	981433.95	.13	.21	.34	212.1	210.7	211.3	48331
206	51 51.78	176 40.55	207.59	981432.36	.20	.20	.40	212.3	210.7	211.3	48072
207	51 51.68	176 40.65	215.47	981432.31	.13	.21	.34	212.9	211.2	211.9	48506
208	51 51.57	176 40.77	240.99	981430.66	.16	.20	.37	213.2	211.3	212.0	47995
209	51 51.37	176 40.73	232.14	981431.28	.20	.19	.40	213.5	211.7	212.4	48326
210	51 51.27	176 40.72	283.59	981428.35	.21	.19	.40	214.3	212.0	212.9	48236
211	51 51.15	176 40.68	284.61	981428.91	.28	.20	.48	215.2	212.9	213.8	48434
212	51 51.02	176 40.45	248.43	981431.81	.91	.22	1.13	216.4	214.7	215.4	48385
213	51 51.02	176 40.28	288.67	981432.84	.66	.22	.88	216.2	214.5	215.2	48126
214	51 50.92	176 40.12	302.67	981428.86	.52	.20	.72	216.9	214.6	215.5	48232
215	51 50.80	176 40.12	309.45	981428.50	.63	.18	.81	217.3	214.9	215.9	48527
216	51 50.72	176 39.98	237.56	981433.29	.55	.21	.75	217.2	215.4	216.2	48512
217	51 50.68	176 39.82	169.17	981437.55	.58	.24	.82	216.9	215.7	216.2	48664
218	51 50.73	176 39.67	96.11	981442.23	.51	.29	.80	216.5	215 9	216.2	48657
219	51 51.72	176 40.98	282.21	981427.85	.11	.20	.30	212.9	210.6	211.5	48167
220	51 51.80	176 41.22	285.31	981427.89	.08	.20	.28	213.0	210.7	211.6	48356
221	51 51.87	176 41.20	281.85	981427.67	.06	.20	.26	212.5	210.1	211.1	48229
222	51 52.10	176 41.18	293.00	981426.80	.08	.24	.32	212.1	209.7	210.6	48174
223	51 52.22	176 41.18	297.84	981426.50	.10	.24	.34	211.9	209.5	210.5	48524
224	51 52.33	176 41.12	328.37	981424.58	.13	.24	.37	212.0	209.3	210.4	48457
225	51 50.83	176 39.55	76.54	981443.19	.53	.29	.82	216.0	215.6	215.8	48415
226	51 50.95	176 39.42	26.44	981446.50	.50	.33	.83	215.7	215.7	215.7	48879
227	51 51.10	176 39.35	21.72	981446.34	.38	.31	.70	214.9	214.9	214.9	48902
A5	51 51.13	176 39.40	19.46	981446.33	.48	.31	.80	214.7	214.9	214.8	. . .
228	51 51.28	176 39.47	27.36	981445.24	.51	.29	.81	214.0	214.0	214.0	48992
229	51 51.40	176 39.42	23.21	981445.19	.23	.28	.51	213.2	213.2	213.2	48403
230	51 51.52	176 39.30	16.30	981445.35	.23	.29	.52	212.7	212.7	212.7	48199
231	51 57.37	176 35.98	213.57	981413.98	.16	.18	.34	186.1	184.4	185.1	48499
232	51 57.52	176 35.83	236.15	981411.99	.26	.18	.44	185.6	183.7	184.4	48486
233	51 57.75	176 35.97	279.20	981408.73	.39	.20	.59	185.0	182.8	183.7	48374
234	51 57.90	176 35.92	337.81	981404.50	.71	.19	.90	184.9	182.3	183.4	48526
235	51 51.02	176 39.17	16.13	981446.91	.69	.32	1.01	215.5	215.7	215.6	48410
FB2	51 51.05	176 39.07	17.65	981446.70	.67	.31	.98	215.3	215.5	215.4	. . .
236	51 51.10	176 38.95	50.75	981444.63	.48	.27	.75	215.2	215.0	215.1	48408
FB3	51 51.12	176 38.98	47.06	981439.15	.44	.27	.72	209.4	209.2	209.3	. . .
237	51 50.98	176 38.62	110.58	981440.98	.91	.23	1.14	216.2	215.7	215.9	48473
238	51 50.97	176 38.28	111.02	981441.63	.33	22	.55	216.3	215.6	215.9	48714
239	51 50.95	176 38.07	171.63	981437.34	.27	.19	.46	216.1	214.8	215.3	48565
240	51 50.90	176 37.75	180.32	981437.80	.22	.19	.41	217.2	215.8	216.4	48145
241	51 50.75	176 37.67	267.56	981432.96	.61	.17	.78	218.9	216.9	217.7	48624
242	51 50.68	176 37.52	160.64	981439.43	.30	.20	.50	217.9	216.7	217.2	48424
243	51 50.50	176 37.38	198.84	981443.05	.39	.24	.63	218 4	217 6	217.9	48721
244	51 50.33	176 37.37	114.04	981442.42	.39	.26	.65	218.3	217.6	217.9	49217

See footnotes at end of table.

Station ID	Lattitude[a]		Longitude[a]		Elevation, ft	Gravity,[b] mgal	Terr. corr.,[c] 2.0 gm/cc			Complete Bouguer anomalies,[d] gm/cc			Corrected magnetics, gammas
							Inner zone	Outer zone	Total	2.00	2.67	2.40	
245	51	50.27	176	37.17	78.45	981444.96	.76	.28	1.04	218.9	218.6	218.7	49376
246	51	50.13	176	37.27	14.44	981449.53	.76	.35	1.11	219.4	219.6	219.5	49183
247	51	50.10	176	37.53	59.15	981446.60	.68	.33	1.01	219.5	219.3	219.4	50117
248	51	49.93	176	37.63	9.12	981449.26	1.32	.39	1.71	219.6	220.1	219.9	49184

[a]Latitude and longitude from state-plane coordinates, AK Zone 10.

[b]Drift-corrected observed gravity

[c]Terrain correction

[d]Bouguer anomalies use post-1967 formula for latitude corrections.

Appendix E

MERCURY CONTENT OF SOIL SAMPLES, NORTHERN ADAK

Sample no.	Location (lat., long.)	Type of soil	Mercury content, ppb
WA-1-82	51°59'40",176°36'40"	Ashy	28
WA-2-82	51°59'31",176°36'30"	Ashy	22
WA-3-82	51°59'20",176°36'25"	Clayey	48
WA-4-82	51°59'05",176°36'40"	Clayey	110
WA-5-82	51°58'35",176°36'30"	Clayey	179
WA-6-82	51°58'20",176°36'30"	Clayey	28
WA-7-82	51°57'50",176°36'30"	Gravelly	33
WA-8-82	51°57'25",176°36'30"	Clayey	80
WA-9-82	51°57'15",176°39'45"	Sandy	42
WA-10-82	51°56'45",176°39'50"	Clayey	107
WA-11-82	51°56'25",176°39'45"	Clayey	151
WA-12-82	51°56'20",176°38'50"	Sandy	129
WA-13-82	51°56'10",176°38'40"		67
WA-14-82	51°55'30",176°38'25"	Sandy	85
WA-15-82	51°57'10",176°39'20"	Sandy	13
WA-16-82	51°57'20",176°38'35"	Sandy	36
WA-17-82	51°57'35",176°37'50"	Sandy	94
WA-18-82	51°57'20",176°37'20"	Rocky	33
WA-19-82	51°57'10",176°36'40"		92
WA-20-82	51°57'15",176°36'05"		73
WA-2i-82	51°57'35",176°35'50"	Clayey	97
WA-22-82	51°56'45",176°35'25"	Clayey	62
WA-23-82	51°57'15",176°34'35"		190
WA-24-82	51°56'40",176°35'15"	Mucky	162
WA-25-82	51°56'50",176°36'05"	Ashy	0
WA-26-82	51°56'35",176°36'45"	Ashy	62
WA-27-82	51°56'20",176°36'15"	Ashy	117
WA-28-82	51°56'10",176°35'20"	Clayey	69
WA-29-82	51°55'55",176°35'15"	Organic mucky	26
WA-30-82	51°55'35",176°35'10"	Ashy	52
WA-31-82	51°54'55",176°35'00"	Sandy	2

Sample no.	Location (lat., long.)	Type of soil	Mercury content, ppb
WA-32-82	51°55'10",176°34'20"	Sandy	8
WA-33-82	51°55'20",176°33'45"	Mucky	111
WA-34-82	51°55'00",176°34'05"	Sandy	24
WA-35-82	51°55'05",176°33'30"	Organic mucky	63
WA-36-82	51°55'10",176°33'10"	Sandy	41
WA-37-82	51°54'40",176°35'15"	Gravelly	19
WA-38-82	51°54'40",176°36'15"	Sandy	44
WA-39-82	51°54'30",176°36'50"	Clayey	152
WA-40-82	51°54'30",176°37'05"	Sandy	31
WA-41-82	51°53'25",176°37'25"	Sandy	39
WA-42-82	51°52'40",176°37'30"	Sandy	6
WA-43-82	51°51'20",176°39'25"	Sandy	27
WA-44-82	51°51'10",176°39'25"	Clayey	111
WA-45-82	51°51'05",176°38'50"	Clayey	58
WA-46-82	51°50'50",176°38'15"	Clayey	44
WA-47-82	51°50'55",176°37'45"	Clayey	100
WA-48-82	51°50'25",176°37'30"	Clayey	62
WA-49-82	51°50'15",176°37'10"	Clayey	35
WA-50-82	51°49'55",176°37'45"	Clayey	107
WA-51-82	51°50'35",176°39'15"	Clayey	27
WA-52-82	51°55'00",176°36'30"	Mucky	98
WA-53-82	51°54'40",176°37'45"	Mucky	109
WA-54-82	51°54'50",176°36'50"		116
WA-55-82	51°54'45",176°37'05"		15
WA-56-82	51°54'35",176°37'55"		109
WA-57-82	51°50'25",176°39'45"		60
WA-58-82	51°50'50",176°40'10"		29
WA-59-82	51°54'40",176°38'30"		96
WA-60-82	51°55'00",176°38'25"	Gravelly	49
WA-61-82	51°55'50",176°39'25"	Organic	107
WA-62-82	51°51'40",176°39'20"	Clayey	13
WA-63-82	51°52'45",176°39'25"	Clayey	62
WA-64-82	51°53'05",176°39'20"	Clayey	51
WA-65-82	51°53'25",176°39'00"	Gravelly	25
WA-66-82	51°53'40",176°38'50"	Clayey	84
WA-67-82	51°54'05",176°39'05"	Clayey	52
WA-68-82	51°54'10",176°38'55"	Clayey	58
WA-69-82	51°53'55",176°39'05"	Clayey	115
WA-70-82	51°53'45",176°39'30"	Clayey	23
WA-71-82	51°53'50",176°40'05"	Clayey	56
WA-72-82	51°51'00",176°40'30"	Clayey	19
WA-73-82	51°51'50",176°40'15"	Organic	88